Springer Textbooks in Earth Sciences, Geography and Environment

AF873239

The Springer Textbooks series publishes a broad portfolio of textbooks on Earth Sciences, Geography and Environmental Science. Springer textbooks provide comprehensive introductions as well as in-depth knowledge for advanced studies. A clear, reader-friendly layout and features such as end-of-chapter summaries, work examples, exercises, and glossaries help the reader to access the subject. Springer textbooks are essential for students, researchers and applied scientists.

Łukasz Pawlik

R Applications in Earth Sciences

From Soil Data to Climate Time Series Analysis and Modeling

Springer

Łukasz Pawlik
Institute of Earth Sciences
University of Silesia
Sosnowiec, Poland

ISSN 2510-1307 ISSN 2510-1315 (electronic)
Springer Textbooks in Earth Sciences, Geography and Environment
ISBN 978-3-031-89675-0 ISBN 978-3-031-89673-6 (eBook)
https://doi.org/10.1007/978-3-031-89673-6

© The Editor(s) (if applicable) and The Author(s), under exclusive license to Springer Nature Switzerland AG 2025

This work is subject to copyright. All rights are solely and exclusively licensed by the Publisher, whether the whole or part of the material is concerned, specifically the rights of translation, reprinting, reuse of illustrations, recitation, broadcasting, reproduction on microfilms or in any other physical way, and transmission or information storage and retrieval, electronic adaptation, computer software, or by similar or dissimilar methodology now known or hereafter developed.
The use of general descriptive names, registered names, trademarks, service marks, etc. in this publication does not imply, even in the absence of a specific statement, that such names are exempt from the relevant protective laws and regulations and therefore free for general use.
The publisher, the authors and the editors are safe to assume that the advice and information in this book are believed to be true and accurate at the date of publication. Neither the publisher nor the authors or the editors give a warranty, expressed or implied, with respect to the material contained herein or for any errors or omissions that may have been made. The publisher remains neutral with regard to jurisdictional claims in published maps and institutional affiliations.

This Springer imprint is published by the registered company Springer Nature Switzerland AG
The registered company address is: Gewerbestrasse 11, 6330 Cham, Switzerland

If disposing of this product, please recycle the paper.

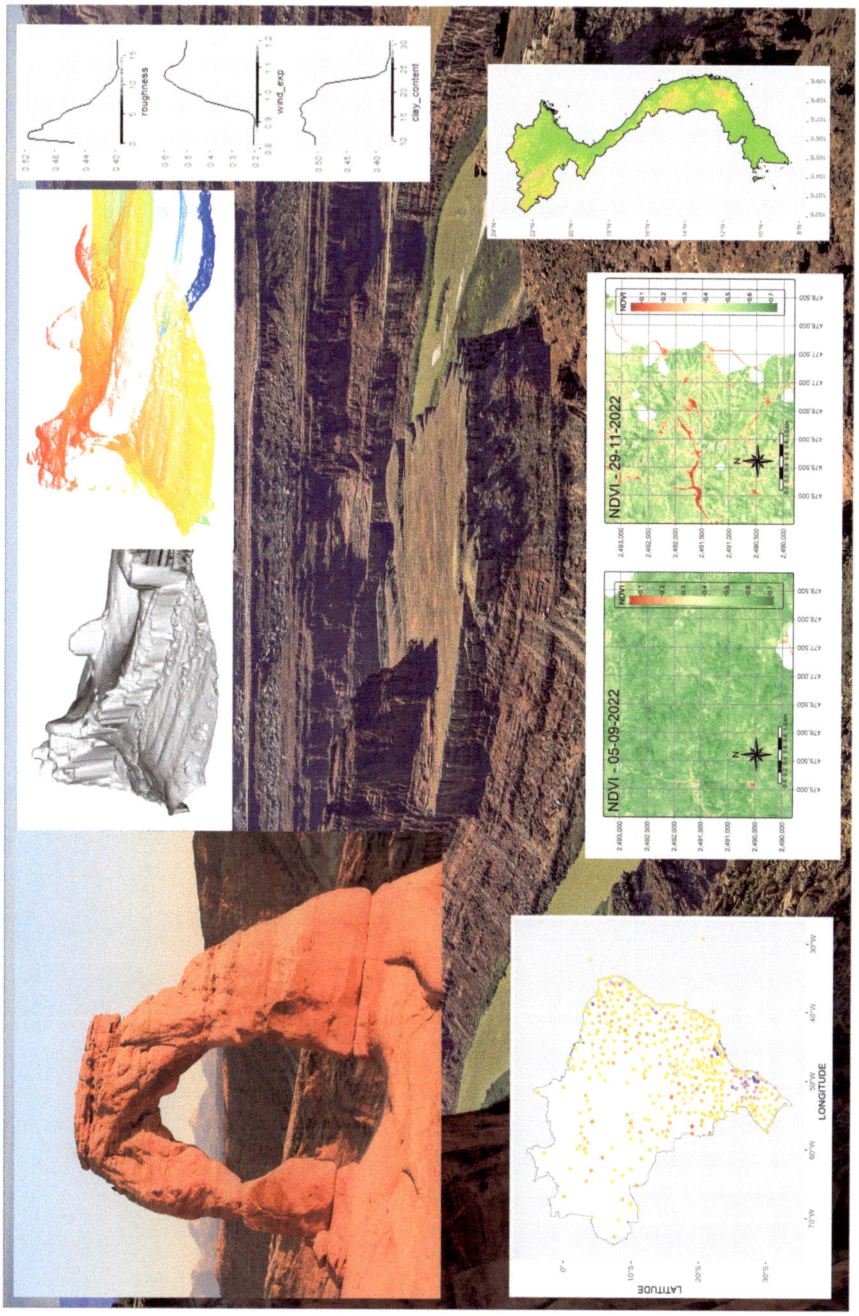

roughness
wind_exp
clay_content
NDVI - 05-09-2022
NDVI - 29-11-2022
LATITUDE
LONGITUDE

For my wife Teresa,
my daughters Aniela, Małgorzata and Weronika,
and my son Tomasz

Preface

Writing the book was a perfect time to reconsider how the data is created and what information is hidden and could be discovered by various available tools and techniques. "Open data" was the phrase frequently used in this context. Luckily, I had a chance to observe the transition from "closed" to "open" data, with the latter expanding rapidly in volume and scope. The benefits of the "open data" and, equally important, "open-source software" are apparent and can lead to extraordinary results at no or minimal cost. Throughout the book, I attempted to show how data can be obtained, analyzed, and displayed using the R programming language. I focused on geoscientific data, which, across many subdisciplines covered under the hood of Earth sciences, have multiple flavors and properties that require a dedicated approach to effectively and correctly explore a given dataset. The book describes various processes studied by geoscientists. However, the approaches presented in each chapter can be easily modified and applied to other data types. The possibility to modify the code for slightly different or new applications is a huge positive aspect of analyses conducted using a scripting language, i.e., all analyses can be repeated, fully understood, and adapted to new problems. I hope this book will help the reader to take the first steps into data mining and visualization using R, Earth-related data, and multiple extra packages that expand the R functionality into many study fields. There is much more to learn, and we can also contribute to the worldwide datasphere with great support from open-source software like R through freely available data and a vast community of developers and users proposing new ideas and solutions.

Sosnowiec, Poland

Łukasz Pawlik

Acknowledgments

I thank my family, my wife Teresa, daughters Aniela, Małgorzata, Weronika, and son Tomasz, for standing behind me and supporting my work on each step of my scientific career. Without you, my work and passion for science would be meaningless. I want to thank the scientific giants and the supervisors I met and worked with at various stages of my education and work. You have shown me the essence of scientific investigation, how to derive knowledge from the data, and how to structure the results into a scientific paper, project proposal, or conference presentation. I will never forget your influence on my career. I want to thank my colleagues, team members, co-authors, and PhD students who pushed my scientific activity to the limit and helped me become a better scientist and academic teacher. I want to thank the reviewers of this book, Prof. Marcin K. Dyderski from the Institute of Dendrology of the Polish Academy of Science in Kórnik, Poland, and Prof. Ola Fredin from the Department of Geosciences of the Norwegian University of Science and Technology in Trondheim, Norway. With your critical evaluation, insightful comments, and suggestions, this book is now in better shape and quality and ready to use by readers of various backgrounds. Thank you for your time.

I want to thank the granting agencies that have supported my research for over a decade. Without their funding, this book would not have been completed. I thank Polish National Science Centre (https://ncn.gov.pl/) for Sonata (2014/15/D/ST10/04123), Preludium-BIS (2019/35/O/ST10/00032), Uwertura (2018/28/U/ST10/00075) and Opus (2019/33/B/ST10/01009; 2023/49/B/ST10/02879) grants, and the National Agency for Academic Exchange (https://nawa.gov.pl/) for Bekker scholarships (PPN/BEK/2018/1/00007; BPN/BEK/2023/1/00055). The University of Silesia in Katowice is thanked for supporting my research through the Research Excellence Initiative.

Competing Interests The author has no competing interests to declare that are relevant to the content of this manuscript.

Contents

Introduction

1

She bought the cheapest train ticket to Paris and began a thousand-mile journey into the unknown.
Obsessive genius, Barbara Goldsmith (2005)

1.1 Data and How They Explain Global Complexity

If the available data are accurate, they may contain hidden (unexplained and uninterpreted) information that can be deciphered and expressed formally (mathematically) in several ways. An in-depth analysis of data can explain, to a certain extent, how Earth system components work and evolve in time. However, raw observations can be noisy and difficult to read and interpret without *a priori* knowledge of the problem and adequate tools and methods. These challenges are common limiting factors and can cause frustration for even highly experienced data analysts. After data preprocessing, validation, and modeling, an initial interpretation of the data can be performed, and conclusions can be formulated. However, the virtual distance covered by the investigation path, from the stage of fieldwork or laboratory experiments to the final results, can be very long, with up to several hours or days spent on data collection, examination, clearance, preprocessing, transformation, and modeling. Some of these steps can be time-consuming and require substantial computational power and computer memory capacity.

Systematic information (datasets and data arrays) for science and industry is similar to oxygen for living organisms. Modern science stems from pure curiosity and necessity ("*our need will be the real creator*"—Plato, 375 BC), which drives the next stages of scientific development (Robertson 1926). Today, scientific institutions, large companies, and other entities cannot function or progress without proper information obtained from large datasets. They rely on this information and plan future steps on the basis of conclusions from it. In the early stage of scientific exploration, the need to decipher and explain unknown phenomena was satisfied by simple descriptions and qualitative data, such as geological formations or landforms (Gilbert 1880). Later, the selection of specific problems, isolated from larger subjects of investigation, was an outcome of increasing problem complexity; notably, when divided into smaller problems, these broader processes were easier to investigate. This natural division resulted in the foundation and development of many new branches of science. In addition, the need for detailed, objective, and systematic information was the main driving force of new formal measurement and monitoring methods given certain rules, patterns, and schemes. Several clear examples include mapping Earth's surface (military topographical maps) and oceans for navigation purposes. However, even before the era of humans conquest of the oceans and new continents, ancient tribes and societies needed to visualize their positions and the surrounding natural or artificial landscape. The creation of simple maps predates the human ability to write (Clarke 2013). The oldest map might be an engraved stone block from Abauntz Cave in Spain dating to 13,660 calBP (Utrilla et al. 2009). The world's oldest geological map is probably the Turin Papyrus from Egypt, dated to 1150 BC (Harrell and Brown 1992). Meteorological observations are another example of the desire of humans to understand the past and perhaps the future. These methods are important for understanding climate conditions and weather forecasting. These data, for instance, are collected and published as part of the Global Surface Summary of the Day (GSOD, provided by the National Centers for Environmental Information, https://www.ncei.noaa.gov/access/metadata/landing-page/bin/iso?id=gov.noaa.ncdc:C00516), which includes daily observations for over 9000 weather stations across the world (Fig. 1.1). However, the first attempts at data acquisition in a systematic manner were likely made by Egyptians as early as

Supplementary Information The online version contains supplementary material available at https://doi.org/10.1007/978-3-031-89673-6_1.

© The Author(s), under exclusive license to Springer Nature Switzerland AG 2025

Ł. Pawlik, *R Applications in Earth Sciences*, Springer Textbooks in Earth Sciences, Geography and Environment,
https://doi.org/10.1007/978-3-031-89673-6_1

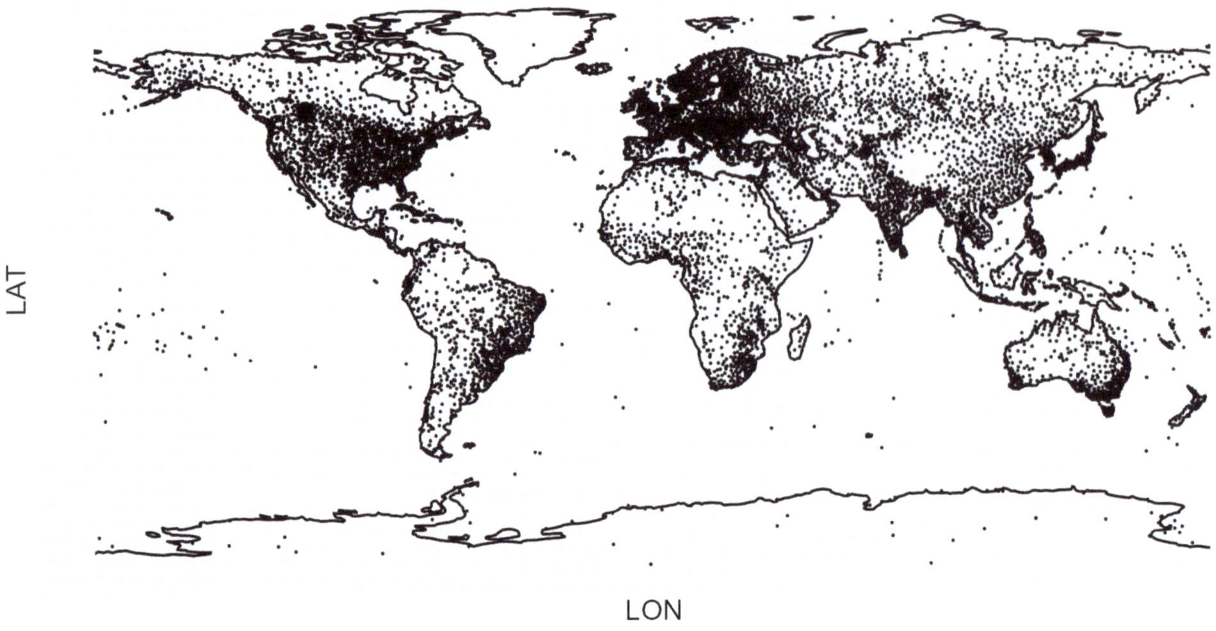

Fig. 1.1 Displaying point data. Weather stations around the world. Information from these stations is available through the Global Surface Summary of the Day

the third century BC. Information on Nile River flood seasonality (measured by the so-called nilometer) is highly important regionally and for the survival of the Ancient Egyptian Empire. Another impressive example of a dataset covering the last 1000 years is the phenological data series of cherry trees flowering in Kyoto, Japan (Aono and Kazui 2008; Aono and Saito 2010). This information was deduced from old diaries and records. In modern times, the Earth's atmosphere, land, and oceans are scanned by satellites. This large stream of remote sensing data has shifted people's understanding of the planet Earth and has aided in identifying dramatic and ongoing changes in terrestrial and marine environments, with deforestation as one of the most striking examples (Hansen et al. 2013).

Box 1.1 Data in a Nutshell

Data for Fig. 1.1. The first layer shows generalized coastlines globally. It is a simple feature collection consisting of several (n = 134) lines (*linestrings,* i.e., vectors linking points and forming lines, as one of the vector object types used in Geographical Information Systems; other vector object types include points, and polygons), with longitude and latitude geographical coordinates expressed in decimal degrees. The layer is defined in the World Geodetic System 1984 (WGS84; a reference system used by the Global Positioning System) coordinate reference system (CRS), which is based on a model of the Earth's shape in the form of an ellipsoid on which the Earth's surface is projected following mathematical transformation (https://earth-info.nga.mil/?dir=wgs84&action=wgs84). Information regarding the reference coordinate system can be obtained by simply typing the name of the defined variable "coastline" in the RStudio console. The second layer is a collection of points representing individual climate stations from which information is regularly transferred to the GSOD database. Access to this database is possible thanks to the **GSODR** package (Sparks et al. 2024). Note that with the **ggplot2** R package (Wickham 2016), two different types of data can be combined in one map. The first type includes spatial objects with an attribute table and can be displayed with the ***ggplot2::geom_sf()*** function. The second type is tabular data with 128,144 rows and 24 columns. We filtered the data to remove redundant information. Note that for each row of data, the information is repeated if it is from the same station. To prepare Fig. 1.1, we only need the station name and its geographical coordinates. A similar map as displayed in Fig. 1.1 can be made with the **leaflet** package (Cheng et al. 2023), which allows us to create dynamic maps in the form of a web-based map application (Fig. 1.2). Recently, the **tmap** package in R has been widely used for map construction (Tennekes 2018).

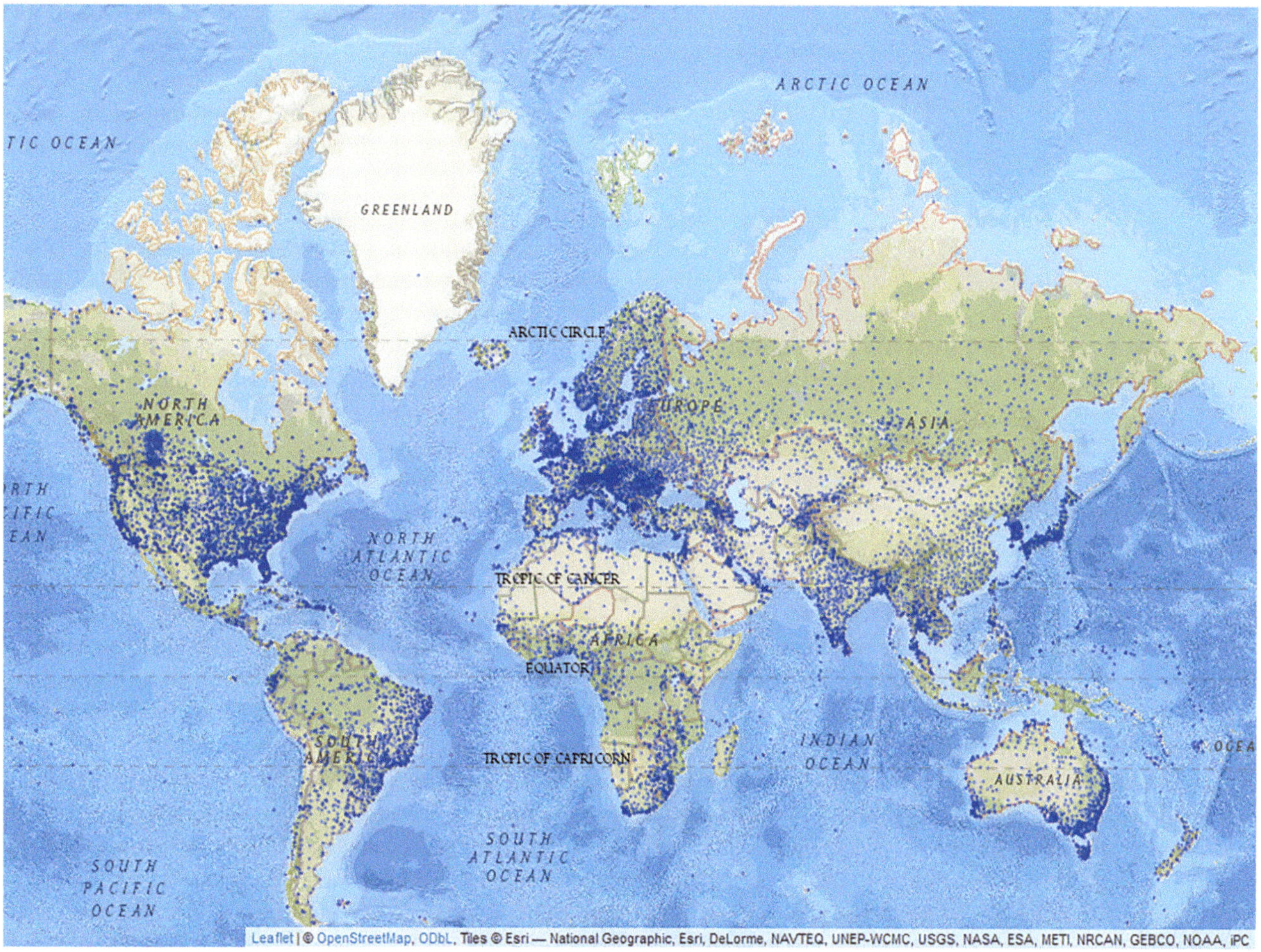

Fig. 1.2 Map containing the same set of information as the map in Fig. 1.1 but with a different background and using the functionality of the **leaflet** package

```
# the pacman::p_load() function is called
before each code chunk is evaluated. It checks
if packages used in the following parts of the
script are installed. If not, it attempts to
install them.
if(!require("pacman")) install.
packages("pacman")
pacman::p_load("GSODR", "ggplot2",
"rnaturalearth", "dplyr")
library(GSODR) #data download from Global
Surface Summary of the Day (GSOD)
library(ggplot2) #data plotting
library(rnaturalearth) #access to spatial data
from https://www.naturalearthdata.com
library(dplyr) #package for data manipulation
#Import a vector layer with coastlines.
coaslines <- ne_coastline(returnclass = "sf")
#Import the geographical positions of
meteorological stations.
stations <- get_inventory()
#Duplicates can be removed by using the
dplyr::distinct() function.
station_names <- stations |>
 distinct(NAME, LAT, LON)
#Plot both datasets on one map.
ggplot()+
 geom_point(data = station_names, aes(x=LON,
 y=LAT), size=0.01)+
 geom_sf(data=coaslines)
#Saving the plot.
#ggsave("fig1_GSOD_stations.png")
pacman::p_load("leaflet", "mapview")
library(leaflet) #supporting interactive Web
Maps preparation
library(mapview) #interactive viewing of
spatial data in R
#Interactive map with climate stations across
the World.
map1 <- leaflet(station_names) %>% setView(lng =
20, lat = 50, zoom=2) %>%
```

```
 addTiles() %>%
 addProviderTiles(providers$Esri.
 NatGeoWorldMap) %>%
 addCircles(lng = ~LON, lat = ~LAT, weight = 1)
#Display the map.
map 1
#Save the map.
#mapshot(map1, file = "fig_leaflet_map.png")
```

1.2 Datasphere, Big and Massive Data, Data Currents, and Streaming

The amount of data produced daily is vast and requires a constant increase in the data storage capacity of computer facilities. One example is satellite images, which are collected every fixed number of days for the same location. The European Space Agency's Sentinel-2 Mission has a 5-day revisit period, producing more than 4 terabytes (TB, equal to 4000 GBs or 4×10^{12} bytes) of new products daily via the Copernicus portal (https://sentinels.copernicus.eu). We now live in the ***Zettabyte Era,*** which likely began in 2010 (Hormann and Campbell 2014), when the total volume of digital data exceeded a zettabyte (10^{21} bytes) for the first time in human history. It was claimed that the so-called digital universe is doubling in size every two years (Lisiecki and Król 2020). In addition, it was estimated that the global dataset will reach 175 zettabytes by 2025, and 46% of these data will be stored in public cloud environments (Reinsel et al. 2018).

The term ***big data*** can be understood as data characterized by considerable variety available in ever-increasing volumes and at constantly increasing rates (or velocities) (https://www.oracle.com/big-data/what-is-big-data/). This definition follows three **V** concepts: *volume, velocity, and variety*. The growing volume of data is an outcome of the invention of modern measurement and monitoring techniques and technology, which allow the capture of a massive variety of information. It is now well understood that every user of the internet produces data, with a so-called digital footprint, which is often used by global companies. In addition, modern observational techniques and highly sophisticated scientific devices produce and store enormous volumes of data, especially in branches of science such as physics, remote sensing, and genetics. All these data sources contribute to ***scientific big data***, which is characterized by complexity, comprehensiveness, and, sometimes, global coverage (Guo 2017). Scientific big data represent natural objects and processes, but the data update rate can differ greatly among disciplines, from a fast update rate for Earth observation data to a slow update rate for certain geological processes (Guo 2017). Currently, data can be analyzed at higher speeds and in larger batches than ever before owing to the development of modern processors, increasing memory capacities, improved algorithms, and parallel computing within supercomputer clusters (high-performance computing (HPC); Navaux et al. 2023). All the data that are produced, used, and stored by users globally form the ***datasphere***.

The continuity of processes and changes in natural objects, a fundamental property of nature, has led to the development of new approaches for data collection, such as continuous observations and monitoring. However, with implements in devices and monitoring methods, it remains an illusion that complete information for a process or an object can be obtained. This is partly true for terrestrial objects; however, any natural process is simply too complicated to obtain full knowledge. Additionally, with the increasing volume of datasets, the internal variability of observations, which must be controlled, increases. A dataset with a high number of observations may not be of high quality. Conversely, single observations, although valuable and provide a glimpse of the full scope of a dataset (population), are often not sufficient for obtaining useful generalizations. Gaps in datasets are common and result from measuring device malfunctions, human errors, or random events, leading to missed observations. Because phenomena are associated with dynamic changes and fluctuations, individual or randomly selected observations do not support assessments of the entire spectrum of related behavior. In this case, environmental monitoring and high-volume datasets can provide partial but sound interpretations and conclusions. Systematic monitoring has greatly contributed to the new era of big data, often referred to as "***data explosion***" (Zhu et al. 2009). Big data production and storage has been partly fueled by the introduction of personal computers and data archives with large computational capacities and volumes for data storage. Along with increasing computational speed, data can now be analyzed in detail with machine learning and deep learning algorithms (Dramsch 2020). In the past, since the invention of the printing press in 1440, *"the explosion of information produced by the printing press had done us a world of good, it turned out*" (Silver 2012). The author also mentioned that, "*It had just taken 330 years—and millions dead in battlefields around Europe—for those advantages to take hold.*"

Data currents and **data streaming** are other phenomena that arose from the need for continuous monitoring and processing of a large volume of information or small chunks of data processed in a fast and efficient way. According to Amazon Web Services (AWS), data streaming is a process involving the continuous emission of high-volume data in an incremental manner, aiming for low-latency processing (https://aws.amazon.com/what-is/streaming-data/), whereas digital data streams (DDSs) involve the continuous digital encoding and transmission of data that may be generated by humans or machines (Pigni et al. 2016).

1.3 Data Exploration and Visualization Methods in Geography and Geosciences

An immense variety of geoscience data collected during fieldwork, laboratory experiments and analyses or obtained by applying remote sensing techniques contribute to a wide range of information that can be accessed using multiple statistical exploration approaches. Selecting a method for data acquisition and exploration is motivated and influenced by several factors including, among others, the key research question and hypothesis, the main data features (e.g., the number and completeness of observations, and variables (predictors)), along with their type (e.g., categorical or numerical) and distribution (e.g., normal or exponential) (Hastie et al. 2009). Determining the most important factors is a challenging task when seeking the best solution on the basis of a set of available data. For example, paleoclimate reconstructions are usually based on fossil material excavated as a core and are thus limited to a point but contain information about the past (potentially several thousand years). Examples include layers of ice from ice cores excavated in Greenland and Antarctica, in which one layer of ice is interpreted as being equivalent to one year of formation given the effects of natural processes (Svensson et al. 2013). In this case, a discrete representation of a space (point) has a quasicontinuous data representation along the time axis (Galton 2019). Point clouds from airborne or terrestrial laser scanning (ALS and TLS, respectively), which are especially useful in **geomorphology** and **geomorphometry**, constitute another type of geoscience data (Pike 2000; Heinaro et al. 2021; Godziek and Pawlik 2023). In this case, information on the coverage of the ground surface is recorded at a certain time. This type of dataset contains millions of points (with known geographical coordinates and elevations), which are laser reflections from the ground surface and natural or artificial objects (buildings, plants, etc.). These data are used for detailed assessments and subsequent visualization of topography and land cover (Fig. 1.3). When the resulting information is correctly applied, a user can obtain a spatially

Fig. 1.3 Point clouds from laser scanning. The Delicate Arch, a sandstone arch from Arches National Park, Utah, USA. (**a**) Ground view (Photo: Ł. Pawlik), (**b**) points classified as ground (class 2) with the elevation attribute (color scale not shown) from airborne laser scanning (Source: US Geological Survey), (**c**) digital elevation model (DEM) with a 0.25 × 0.25 m horizontal resolution interpolated from ground points. The black arrows point to the Delicate Arch in the point cloud data (**b**) and the DEM (**c**)

continuous model of the ground surface at a very high horizontal resolution with a selected spatial interpolation algorithm, such as inverse distance weighted (IDW) interpolation (Achilleos 2008).

Box 1.2 Data in a Nutshell

This example is based on a point cloud imported from the United States Geological Survey (USGS) archive. We first define a place for which data is imported. We selected the Delicate Arch in the Arches National Park, Utah. The tile consisting of the area covering the Delicate Arch and its surroundings can be imported manually and stored in a working catalog on a local hard disc, or it can be saved in a temporary file in the RStudio session working memory. Because the volume of the dataset is very large, it can be reduced to the size of a buffer zone extending 100 m from the point of interest. To complete this step, we use the **lidR** package and the ***clip_circle()*** function (Roussel et al. 2020). The data are in LAZ format, which is a compressed LAS file, a standardized binary format for storing 3-dimensional point cloud data and point attributes. For each data point, three coordinates, namely, X (longitude), Y (latitude), and Z (elevation), are measured during airborne laser scanning. Using a direct link to the data that is displayed after clicking the desired tile in the 3DEP LidarExplorer of the USGS (https://apps.nationalmap.gov/lidar-explorer/#/), the data can be downloaded to a personal computer using the ***download.file()*** function in R. The file occupies 552 Mb of disc storage, covers 2.25 km^2, and contains 16 million points, with a resolution of at least 7 points per m^2. The spatial data are in the NAD83/UTM zone 12 N coordinate reference system, which is based on the GRS 1980 ellipsoid. This information can be displayed using the ***crs()*** function. The imported point cloud consists of points classified according to the standard introduced by the American Society for Photogrammetry & Remote Sensing (ASPRS): 0—created, never classified, 1—unclassified, 2—ground, 3—low vegetation, 4—medium vegetation, 5—high vegetation, 6—building, 7—low point (noise), 8—model key-point (mass point), 9—water, 10—reserved for ASPRS definition, 11—reserved for ASPRS definition, 12—overlap points, 13–31—reserved for ASPRS definition (ASPRS 2019). If we use points from class 2, the ground surface can be displayed as a DEM. For this task, the ***rasterize_terrain()*** function with the default argument can be used for spatial interpolation on the basis of "*knnidw*", which stands for the k-nearest neighbor (KNN) with inverse-distance weighting (IDW) method. However, because the Delicate Arch is 16 m in height, it is not correctly classified as a ground surface when using default parameters. This isseu needs to be corrected by applying ***classify_ground()*** function and the ***pmf*** (progressive morphological filter) algorithm (Zhang et al. 2003) with the ***th*** (threshold height) argument set to 20 m. Figure 1.3b shows only ground points, and Fig. 1.3c depicts the interpolated DEM at a horizontal resolution of 0.25 × 0.25 m.

```
pacman::p_load("lidR")
#Data from 3DEP LidarExplorer: https://apps.
nationalmap.gov/lidar-explorer/#/
#Direct link to the dataset. We create this
object and use it in the next step: the data
download and storage in the session's temporary
memory.
url <- "https://rockyweb.usgs.gov/vdelivery/
Datasets/Staged/Elevation/LPC/Projects/UT_FEMA_
FS_FlamingGorge_2020_B20/UT_
FlamingGorge_5_2020/
LAZ/USGS_LPC_UT_FEMA_FS_FlamingGorge_2020_
B20_12SXH300885.laz"
#We create a temporary file.
destfile
<- tempfile(pattern = "USGS_LPC_UT_FEMA_FS_
FlamingGorge_2020_B20_12SXH300885",
                     tmpdir = tempdir(),
                     fileext = ".laz")
#If the Internet connection needs longer than
the default 60 seconds, the timeout can be
changed using the options() function.
options(timeout = 600)
#Download the file.
download.file(url, destfile, mode = "wb")
#Load the library supporting LiDAR data
analyses.
library(lidR)
#Read the data from the temporary file. The
'select' argument specifies the data that will
be loaded into the R session memory. 'xyzcrn'
means that the x, y, and z coordinates,
```

```
classification data, r - return number, and
n - number of returns are loaded.
darch <- readLAS(destfile, select = "xyzcrn")
#The Delicate Arch geographical position in the
WGS84 coordinates.
darch_position <- c(-109.499313, 38.743629)
#Filter ground points and the first returns.
darch_gfr <- filter_poi(darch, Classification ==
2L & ReturnNumber == 1L)
#Reduce the data into a 100 m radius. Note that
coordinates of the 'darch' dataset are in
meters.
darch_clip <- clip_circle(darch_gfr, xcenter =
630418, ycenter = 4289396,
                                radius = 100)
#Interpolate the data.
darch_idw <- rasterize_terrain(darch_clip,
algorithm = knnidw(k = 10L, p = 2),
                                     res = 0.25)
#Plot the digital elevation model.
plot_dtm3d(darch_idw, bg = "white")
#We noted that the Delicate Arch was not
classified as the ground surface. How to fix it?
#We load the data again but without information
on point classification. 'xyzrn' means that the
x, y, and z coordinates, r - return number, and
n - number of returns are included.
darch <- readLAS(destfile, select = "xyzrn")
#We need to repeat this operation because we
have modified the dataset.
darch_clip <- clip_circle(darch, xcenter =
630418, ycenter = 4289396,
                                radius = 100)
#Plot the data.
plot(darch_clip, bg = "white", legend = TRUE)
#Classify points from the point cloud. We did
not load a part of the information on
classification, so now we need to classify these
points.
darch_ground <- classify_ground(darch_clip,
algorithm = pmf(ws = 5, th = 20))
#Again, we filter only the ground points and the
first returns.
darch_gfr1 <- filter_poi(darch_ground,
Classification == 2L & ReturnNumber == 1L)
#Interpolate the data.
darch_idw <- rasterize_terrain(darch_gfr1,
algorithm = knnidw(k = 10L, p = 2),
                                     res = 0.25)
#Plot the digital elevation model.
plot_dtm3d(darch_idw, bg = "white")
```

Next, another important step in an analysis can be initiated. This step is related to the most basic research task in descriptive statistics: assessing the correlation between variables or differences between groups of data. Here, exploratory data analysis (EDA) involves a set of tools. This approach was named by Tukey (1977), who perceived it as a form of investigative work. During EDA, several data exploration techniques are used in as many ways as possible until a plausible relation among the data is found (Yu 2010). ***EDA*** is the first step of data evaluation in terms of structure, completeness, and validity (Vigni et al. 2013).

A closely related EDA approach is ***data mining*** (also known as ***knowledge discovery from data*** (KDD)), which can be defined as the computerized process of extracting previously unknown, interesting, and important information, patterns, and knowledge from a large database (Han et al. 2012; Shi 2014). ***Statistical learning***, which is an approach for understanding data (James et al. 2021), is similar to *data mining*. In statistical learning, approaches for estimating $\boldsymbol{f}$, which is a fixed but unknown function representing the systematic information that $\boldsymbol{X}$ provides about $\boldsymbol{Y}$, are applied according to the general form $\boldsymbol{Y = f(X) + \epsilon}$, where $\boldsymbol{\epsilon}$ is a random error term (James et al. 2021). Knowledge discovery is frequently an iterative process that involves the following sequence: (1) data cleaning, (2) data integration, (3) data selection, (4) data transformation, (5) data mining, (6) pattern evaluation, and (7) knowledge presentation (Han et al. 2012).

An important part of data mining and modeling is data visualization. It is a key component of any data analysis workflow and is essential for successfully disseminating results. Data visualization is the backbone of data science and is applied in various stages of data exploration and modeling. During some cases, correct conclusions can be drawn from data only after the examination of a resulting diagram. In the initial stage of data exploration, plotting the data can help us understand the data itself and its completeness and structure, and it allows the formulation of initial hypotheses. The importance of this step is frequently supported by Anscombe's quartet (Anscombe 1973). Notably, the same linear regression model is constructed on the basis of a different set of observations (Fig. 1.4). The case in which the model best fits the data is then determined, but only after graphical display because the model statistics are very simi-

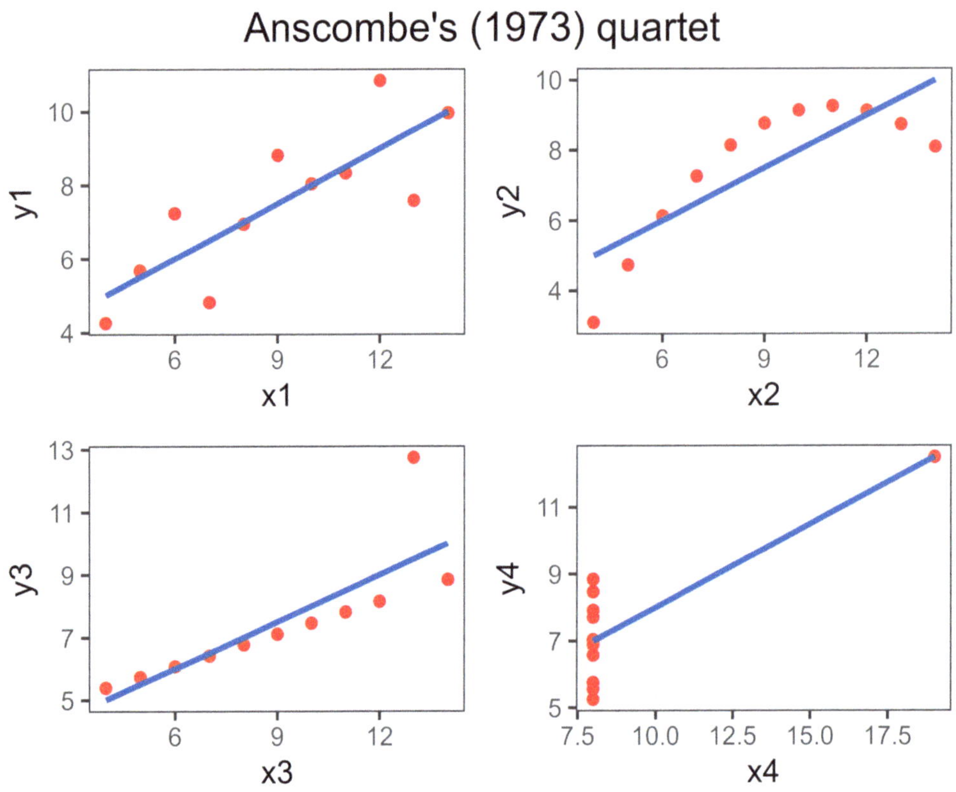

Fig. 1.4 Anscombe's paradox. Illustration of the importance of plotting data on the basis of Anscombe's quartet (Anscombe 1973)

Table 1.1 Summary statistics for Anscombe's quartet model

	Model 1	Model 2	Model 3	Model 4
(Intercept)	3.000 (1.125)	3.001 (1.125)	3.002 (1.124)	3.002 (1.124)
x1	0.500 (0.118)			
x2		0.500 (0.118)		
x3			0.500 (0.118)	
x4				0.500 (0.118)
Num.Obs.	11	11	11	11
R2	0.667	0.666	0.666	0.667
R2 Adj.	0.629	0.629	0.629	0.630
AIC	39.7	39.7	39.7	39.7
BIC	40.9	40.9	40.9	40.9
Log.Lik.	−16.841	−16.846	−16.838	−16.833
F	17.990	17.966	17.972	18.003
RMSE	1.12	1.12	1.12	1.12

lar (Table 1.1). The model formula is $y = 3 + 0.5x$. The upper right panel shows observations that follow a nonlinear model and might capture the relationship between variables y2 and x2 better than can other models. The lower left graph shows one outlier that causes a deviation from the remaining data. Without this outlier (which might be an error or incorrectly typed value), the model would fit the data almost perfectly. The lower right diagram shows an outlier with which all the information about the slope of the regression line is associated (Anscombe 1973).

Box 1.3 Data in a Nutshell

Each data is a mystery until patterns, regularities, and schemes that are hidden within sets of numbers or categories are discovered. However, what if different data can be summarized by exactly the same model but some of the datasets are problematic. Such a situation was described by Francis John Anscombe, who highlighted the use of diagrams as an important source of discoveries and a tool in a statistician's hands. One of his famous statements was "*A computer should make both calculations and graphs*" (Anscombe 1973). In Fig. 1.4, Anscombe's quartet of scatterplots is reproduced with the same linear model. In R, the dataset is included in the **datasets** package and can be used directly. To create a simple visualization, the **ggplot2** package is used with the ***geom_smooth()*** function, which establishes a regression line. Next, we use the **ggpubr** package (Kassambara 2023) and the ***ggarrange()*** function to combine four plots into one figure. The **modelsummary** package is used to combine the statistics for the four models into one table and save it as a PNG file (Arel-Bundock 2022). Note that the ***modelsummary()*** function accepts a list of models and use it to build the table.

```
pacman::p_load("ggplot2", "modelsummary",
"webshot2")
library(ggplot2)
library(modelsummary) #summary tables and plots
for statistical models and data
pl1 <- ggplot(anscombe, aes(x=x1, y=y1))+
 geom_point(color = "red")+
 geom_smooth(method = "lm", se = FALSE)+
 theme_bw()
pl2 <- ggplot(anscombe, aes(x=x2, y=y2))+
 geom_point(color = "red")+
  geom_smooth(method = "lm", se = FALSE)+
  theme_bw()
pl3 <- ggplot(anscombe, aes(x=x3, y=y3))+
  geom_point(color = "red")+
  geom_smooth(method = "lm", se = FALSE)+
  theme_bw()
pl4 <- ggplot(anscombe, aes(x=x4, y=y4))+
  geom_point(color = "red")+
  geom_smooth(method = "lm", se = FALSE)+
  theme_bw()
#Arrange the plots.
library(ggpubr) #package supporting editing and
customization of plots created with the ggplot2
library
pl5 <- ggarrange(pl1, pl2, pl3, pl4)
#Add a title.
annotate_figure(pl5, top = text_grob("Anscombe's
(1973) quartet",
 size = 14))
#Save the graph. When saving the plot locally,
the user has to define the exact file directory
following the pattern "Disc_partition://Folder_
name/File_name.png", for instance, "D://
MyFolder/myplot.png")
#ggsave("anscombe.png", dpi = 600)
#Display model statistics in a table form.
models <- list(
"Model1" = lm(x1~y1, data=anscombe),
"Model2" = lm(x2~y2, data=anscombe),
"Model3" = lm(x3~y3, data=anscombe),
"Model4" = lm(x4~y4, data=anscombe)
)
modelsummary::modelsummary(models, title =
"Anscombe's (1973) quartet models statistics")
#If you want to save it define the output
argument in the modelsummary() function. The
output should be the exact file directory with
the file name and format, e.g., png.
#modelsummary::modelsummary(models, title =
"Anscombe's (1973) quartet models statistics",
output = "anscombe_table.png")
```

Each diagram or chart displays measured quantities, and to pass a message to the reader or communicate with the public, points, lines, a coordinate system, numbers, symbols, words, shading, and color are used (Tufte 2001). One of the pioneers of statistical data visualization was William Playfair (1759–1823), a British engineer, economist, and statistician. In his pioneering studies (Playfair 1786, 1801), he showed how well-designed charts based on statistical data can engage the reader's imagination, catch the audience's attention and convince the public of the value of the underlying piece of information (Fig. 1.5).

Currently, data visualization is highly popular, and many leading international scientific journals are devoted to developing effective and attractive methods of knowledge display in the form of graphs, plots, charts, figures, and graphical abstracts. Several recommendations for graphical display were proposed by Edward R. Tufte to enhance the visualization of data and results (Tufte 2001):

1. Show the data.
2. Encourage the viewer to think about the substance rather than about the methodology.

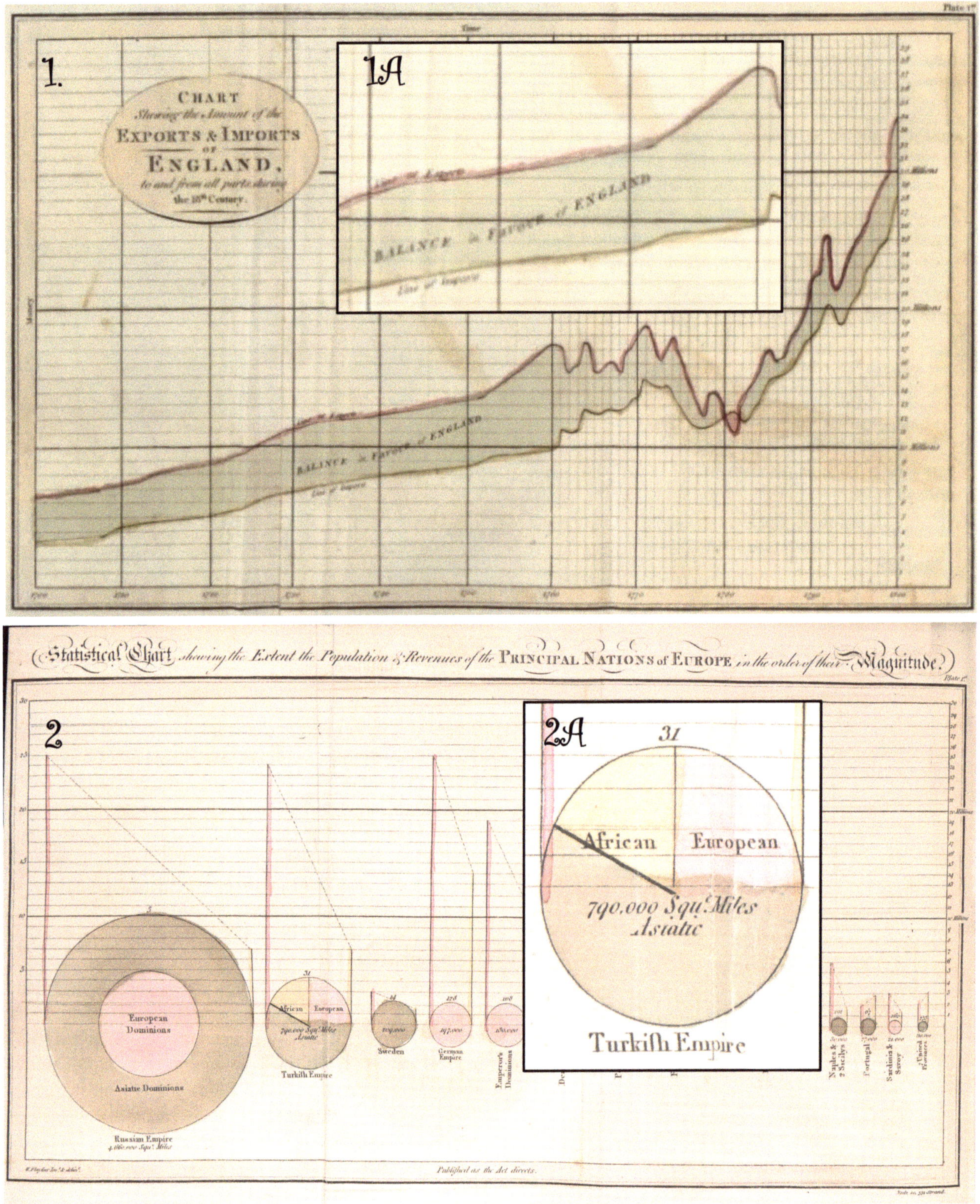

Fig. 1.5 William Playfair's charts illustrating (1) the exports and imports of England in the eighteenth century (in millions of pounds). The upper chart is the first plate of his famous Commercial and Political Atlas (Playfair 1786). Imports are marked in yellow, and exports are marked in red. The area of the difference between imports and exports is shaded to stand out from the background (1A). Notably, the part of the graph with higher changes corresponds to denser coordinate lines, which help readers find the correct values on the X- and Y-axes. The lower chart is from The Statistical Breviary (Playfair 1801) and shows four pieces of information (2): countries as pie charts, countries geographic positions in reference to continents (pink—Europe, yellow—Africa, brown—Asia), country area (square miles), and population in millions (Y-axis). Subplot 2A shows that the Turkish Empire occupied parts of three continents in the eighteenth century

3. Avoid distorting what the data have to say: "*Data graphics should draw the viewer's attention to the sense and substance of the data*" (Tufte 2001).
4. Make large datasets coherent.
5. Encourage the eye to compare different pieces of data.
6. Reveal the data at several levels of detail.

A plot should attract the attention of the reader and be as informative as possible. A graphical display of the data should be easily understandable without reading the text. More technical recommendations were given by Midway (2020):

1. Diagram first (think about the core message you want to pass to the reader) (Bingham and Fry 2010).
2. Use the right software (use R or Python as often as possible; these are open-source tools that make the results broadly interpretable).
3. Use an effective geometry, and show data (for example, wind speed and direction are better displayed, such as with rose charts, than described with text).
4. Colors always mean something (when colors are integrated with information, such as choropleths and raster maps, color scales should be used (Brewer 2017)).
5. Include uncertainty (without information on uncertainty, the statistical representativeness of results is limited).
6. Panel, when possible (small multiples are useful in a figure to highlight differences).
7. Data and models are different things (e.g., raw data, summarized data, or an inferential statistic).
8. Simple visuals, detailed captions (figures should be self-explanatory, and captions should be standalone).
9. Consider an infographic (they were found to have the highest memorability score in one study (Borkin et al. 2013)).
10. Get an opinion (find an external reviewer for your figure).

1.4 Quantitative Geography: History of the Revolution

In this chapter, we refer to one example of the discipline transformation and advancement associated with the adaptation of data-driven and model-based methods. We highlight ***quantitative geography*** and ***quantitative revolution***, which are considered important parts of Earth's scientific history. Like many scientific disciplines, quantitative geography is based on datasets, data analysis and modeling, and is the backbone of the ***quantitative revolution*** (Burton 1963). As noted by Gould (1979), *the intellectual space called geography has changed and sharply increased in multidimensionality*. This change represents a shift from qualitative descriptions of the Earth's surface and processes acting within various Earth's spheres to quantitative descriptions. Many branches of today's geosciences have benefited from the early development of physical geography and, since the 1940s and 1950s, from quantitative geography and transformations in the physical and biological sciences (Burton 1963). ***Geographic information systems*** (GISs), which ***developed into geographic information science in the 1990s*** (Goodchild 1992), and ***geocomputation*** are strongly rooted in quantitative geography (Yano 2001; Waters 2018). ***Quantitative geography*** is a discipline that focuses on the analysis of numerical spatial data, the development of spatial theory, and the construction and testing of mathematical models of spatial processes (Fotheringham et al. 2000). The predominant focus of spatial data is what distinguishes quantitative geography from other disciplines (Fotheringham et al. 2000), and the primary function of GISs is to make spatially oriented information available in a usable form (Dacey and Marble 1965). However, ***geocomputation*** involves the use of geodata and the development of geotools for improved scientific inference (Lovelace et al. 2019). Here, we cannot forget ***automatic cartography,*** which is an important step in the development of modern GISs and spatial data visualization (Bickmore and Evans 1970). The quantitative revolution and development of quantitative geography would not be possible without advancements in computational speed and capacity (Krumbein 1962; Haggett 1969; Sui and Morrill 2004) and early ground-breaking ideas regarding artificial computing using machines (Turing 1950). Figure 1.6 shows an initial attempt at spatial data visualization; after decades of advancement, this basis was eventually transformed into the digital elevation and terrain models widely used today (Young and Evans 1978).

1.5 Spatial Data, Spatial Data Science, GIS, and Geocomputation

Claudius Ptolemy was the first person who, in the second century A.D., proposed a world map with known, at the time, places expressed in geographical coordinates (latitudes and longitudes). He produced 27 maps of the known world, which included Europe, Asia, and Africa (Baynton-Williams 1969). Since his time, several centuries have passed, and the foundation for spatial analyses has been firmly built on people's desire to understand their surroundings and ground topography. The beginning of spatial analysis was linked to two distinct figures: Valentine Seaman (1770–1817) and John Snow (1813–1858). In 1798, Valentine Seaman published maps illustrating the distribution of yellow fever in

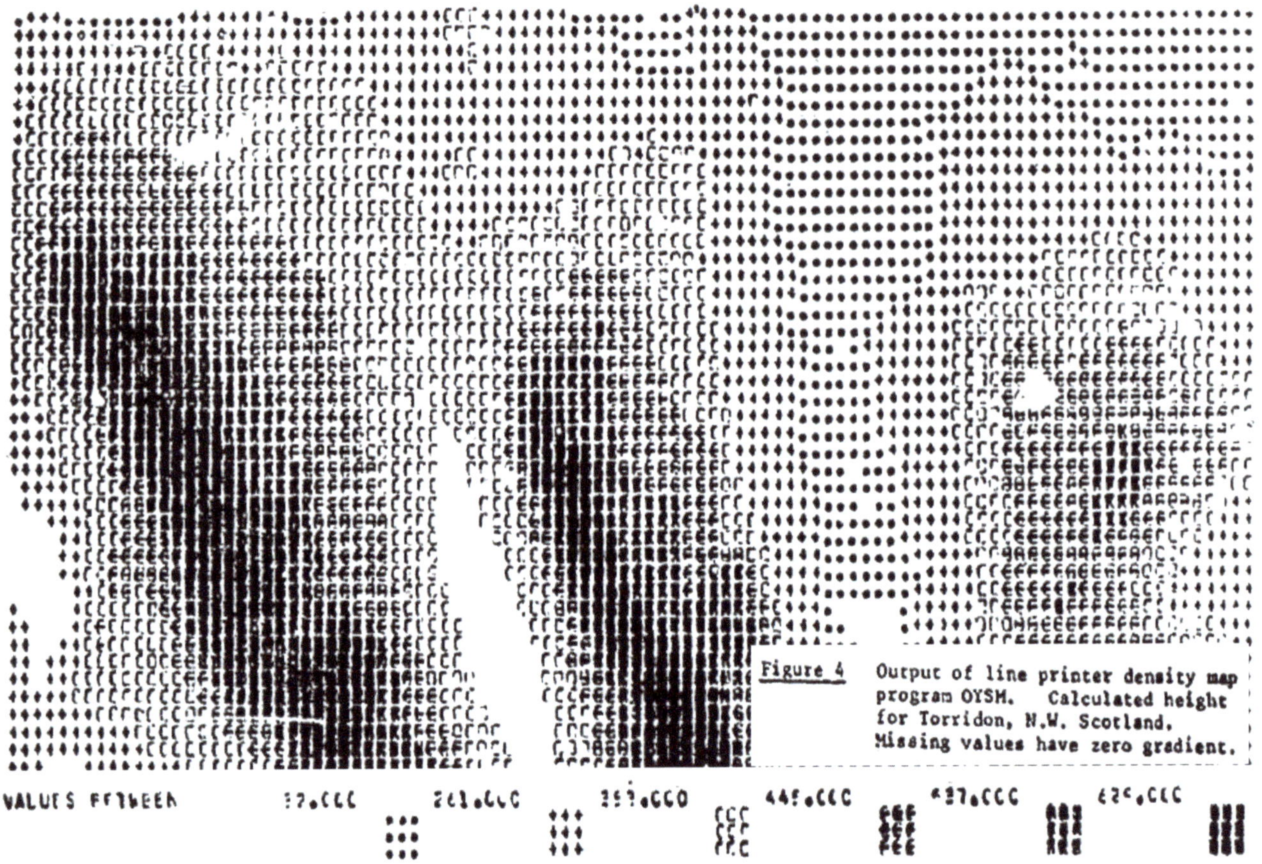

Fig. 1.6 The terrain height above sea level is displayed as labels indicating different elevation classes. This figure was adapted from Young and Evans (1978)

New York (Seaman 1798). These maps predate the more widely known cholera epidemic map of London made by John Snow in 1854, which is a considered pioneering work in the field of spatial analysis (Buzai 2020). John Snow assumed that cholera is spread by contaminated water, and to test this hypothesis, he combined information on cholera deaths and water pumps used to supply water in the area around Broad Street in London, where the number of deaths was unusually high (Snow 1855). Figure 1.7 shows how he assessed the spatial dependency of the two initially not clearly related factors: the source of drinking water and the deaths of Londoners. Through his work, John Snow made an initial contribution to the widely adapted concepts of spatial analysis and location, spatial distribution, association, interaction, and evolution (Buzai 2020).

Spatial data are everywhere (Bivand et al. 2008), and everything seen on a map can be described as geographical information (Galton 2019). ***Spatial data science*** is considered to have grown from many existing scientific disciplines and industrial activities (Pebesma and Bivand 2023). Data, which are a subject of exploration in geography and geosciences, have key attributes, such as known spatial positions expressed by geographical coordinates, namely, longitudes and latitudes (Fig. 1.8). These attributes allow the user to perform spatially oriented analyses and map preparation. One of the main concepts driving spatial data science is the **First Law of Geography**, which states that "*everything is related to everything else, but near things are more related than distant things*" (Tobler 1970). When spatial data are collected repeatedly, ***data cubes, spatiotemporal arrays,*** or ***Earth system data cubes*** (ESDCs) are produced and can be used to interpret given phenomena from long-term and global perspectives (Mahecha et al. 2020; Montero et al. 2023; Pebesma and Bivand 2023). ***Data cubes*** are defined as arrays of data with one or more dimensions associated with space and/or time (Lu et al. 2018). Discrete data points can be the subject of **spatial interpolation,** which results in spatially continuous data expressed by grid cells in raster models, which are commonly used in GISs (Heywood et al. 1998) (see Sect. 1.6). Owing to their known positions, observational and monitoring data can be analyzed even more comprehensively, and spatial relationships (e.g., distance) between different objects can be evaluated and interpreted. Increased precision in geographical positioning is an out-

Fig. 1.7 Two examples of early maps illustrating the first attempts at spatial analysis in health geography and the spread of disease. (1) John Snow's map displaying cholera deaths in 1854 around Broad Street (now Broadwick Street) in London. A water pump with contaminated water was located in the central part of Broad Street. The image (1A) shows the current state of the historical water pump on Broadwick Street in London (Photo: Łukasz Longosz, reproduced with permission). (2) Valentine Seaman's map from 1798 showing the spread of yellow fever on the streets of New York City. The red/yellow dots indicate fatal cases, and "S" (marked with black arrows) represents sites of contagion or effluvium (or other miasma sources) (Altonen 2012)

come of constant improvements in the precision of measuring devices used during ground, aerial, and space measurement campaigns.

Another rich branch of spatial data science and ***thematic cartography*** is related to generalized data (quantitative and qualitative), which are displayed as ***choropleths***. These data do not have an exact geographic position but rather express general properties, tendencies, and facts. Thematic cartography deals mostly with spatial data visualization.

1.6 Spatial Models and Data Interpolation

Spatial models are an important part of spatial science, geography, and Earth science research. The first spatial models, such as von Thünen's land use model and Ravenstein's migration model, which are based on gravity principles, were proposed in the nineteenth century (Fischer and Nijkamp 1993; Cheshire and Kent 2023). Incomplete data, low-quality data, our inability to continuously measure everything everywhere, and the lack of practicality of field measurement campaigns in many areas paved the way for various methods of data ***interpolation*** and ***extrapolation***. ***Interpolation***, in the most common sense, is a mathematical operation aimed at filling gaps between observations to obtain a complete dataset. Incomplete time series or spatial data require a data imputation technique based on different rules that outputs continuous information at a desired time frequency (hours, days, months, or years) or spatial resolution. It is not uncommon that time series suffer from a lack of observations, which can result from device failure or damage. In such cases, the evaluation of the main time series components (trend, cyclicity, or seasonality) becomes difficult. ***Spatial interpolation*** is another key methodology in the geosciences. The main aim of this approach is to obtain continuous surfaces in the form of a ***raster model***. Given a set of discrete points or subareas, spatial interpolation is used to find a function that can represent the whole surface (Siu-Ngan Lam 1983). The procedure involves the prediction of

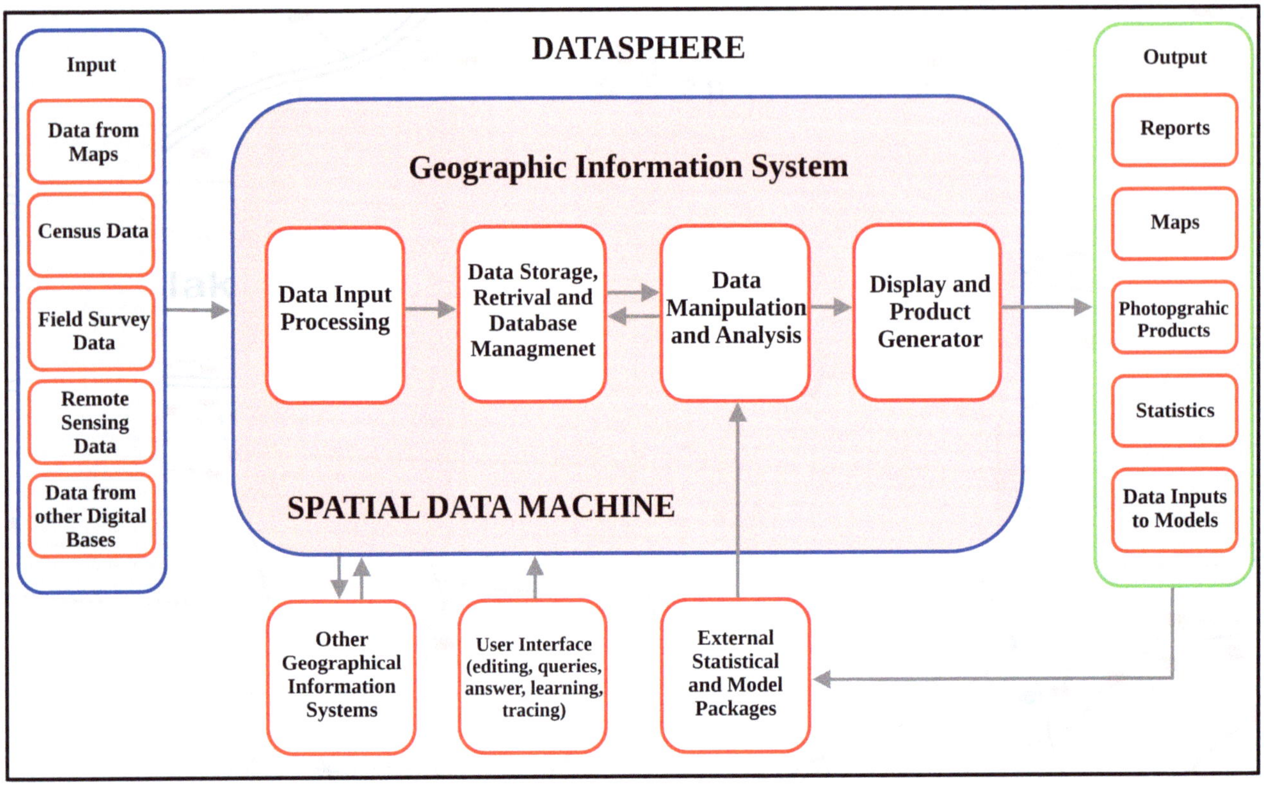

Fig. 1.8 The structure of a geographic information system that relies on various input data, devices, software, and algorithms and has an appropriate storage capacity for producing maps and spatial models. Figure adapted from Fischer and Nijkamp (1993)

feature values in unsampled locations. Interpolation models can also be considered to reflect certain hypotheses about the surface, which may or may not be true (Siu-Ngan Lam 1983). Generally, spatial interpolation methods are classified as nongeostatistical, geostatistical, or combined methods (Li and Heap 2014). Additionally, these methods are further classified into exact or inexact methods and local or global methods. In contrast, ***extrapolation*** is the projection, extension, or expansion of known data or experiences into an unknown area outside the range of the training data (Miller et al. 2004; Kuhn and Johnson 2013).

1.7 Models and Modeling: Simplifying the Data Complexity

A model is any sort of function that has predictive power (Burger 2018). The primary value of models is heuristic (Oreskes et al. 1994). A model represents an attempt to capture the complexity of a real situation, process, or object, i.e., "*an abstraction of reality*" (Demeritt and Wainwright 2005). However, not all models are explanatory (Craver 2006). A model is a simplification of a real relationship or a process developed over time and provides synthetic information that can be easily interpreted (Skogen et al. 2021). However, all models should be treated with caution, and it was claimed that "*Essentially all the models are wrong, but some are useful. However, the approximate nature of the model must always be borne in mind.*" (Box 1976). A map is a classic model of reality (Fig. 1.9). Models can be qualitative (conceptual) and are typically proposed during the first stage of a study (Demeritt and Wainwright 2005; Bokulich and Oreskes 2017). One example is the soil formation model used in Jenny's (Jenny 1941) state factor equation: $s = f(cl, o, r, p, t, \ldots)$ (Gerrard 1992), where s is a considered soil property, cl is the climate, o represents living organisms, r is relief, p is parent material and t is time. This simple formula is an attempt to capture the highly complex multiprocess path of soil formation, which can span thousands of years. Possible? Most likely, yes, but with a high level of generalization. Undoubtedly, any attempt at obtaining a qualitative or mathematical description of a complex real-world process should be considered valuable because we cannot afford to live in an unexplained world. Many times, a lack of information has proven costly. Another example of a conceptual model was the model of the cycle of erosion proposed by William

Fig. 1.9 Map (model) of Europe in the early twentieth century with simplified borders of countries engraved on a soldier's aluminum canteen. This artifact belongs to the Polish Army Museum in Warsaw, Poland. This canteen, identified as model 37, that was produced in 1939 at the "Światowit" plant in Myszków (inventory no. MWP 44060*), Poland. It was donated to the collection of the Polish Army Museum by the Sikorski Institute in London. According to the institute, the canteen belonged to a Polish soldier stationed at the Miranda del Ebro camp, Spain. Photo: Ł. Pawlik (with permission from the museum authorities)

Morris Davis at the end of the nineteenth century (Davis 1899; King and Schumm 1980) (Fig. 1.10). Although widely criticized, this model currently represents one of the classic attempts at conceptualizing landscape evolution (King and Schumm 1980).

Models can be ***physical***, ***numerical***, ***linear***, ***nonlinear***, ***stochastic*** (probabilistic), ***deterministic***, or ***mechanistic***. It is likely that, as with all natural phenomena, no fully ***deterministic model*** (mathematical model based on a set of equations) exists, and an unknown component that disrupts a deterministic model should always be considered (Box et al. 2016). A ***stochastic model*** predicts outcomes weighted on the basis of their likelihoods or probabilities (Pinsky and Karlin 2011). ***Physical models*** are slices of real interactions reconstructed in a laboratory, mimicking real processes on a smaller scale and under controlled conditions. ***Numerical models*** are mathematical models that approximate natural systems and their interactions through a system of equations (Bokulich and Oreskes 2017), where a system is a structured set of objects and/or attributes (Chorley and Kennedy 1971). ***Mathematical models*** consist of simple equations but can take the complex form of software codes (Demeritt and Wainwright 2005). A ***mechanistic model*** captures the mechanistic relationship between inputs and outputs (Baker et al. 2018). It mimics real-life events by using information on underlying mechanisms expressed in simplified mathematical formulations of causal mechanisms (Baker et al. 2018). In this sense, the model is also explanatory (Fig. 1.11). ***Mechanistic models*** are considered complete when they include all relevant features of a mechanism, its component

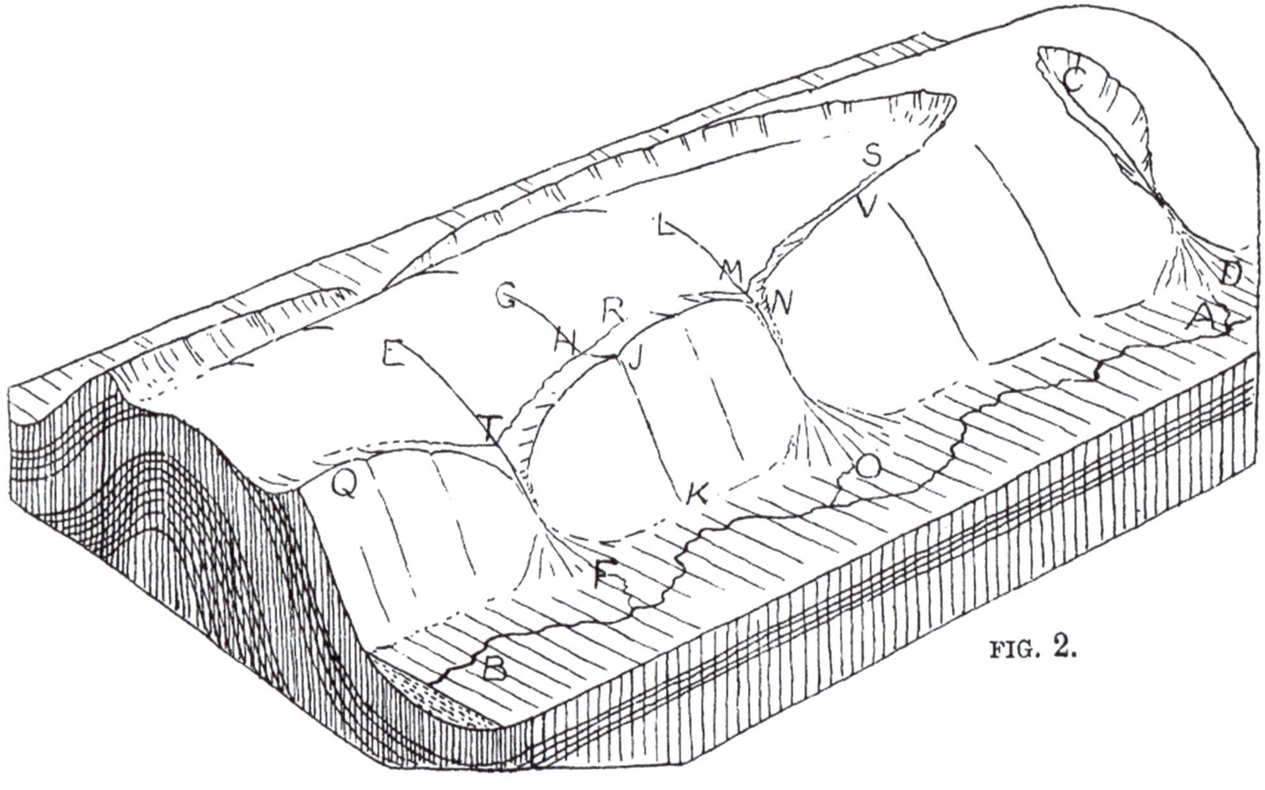

FIG. 2.

Fig. 1.10 A conceptual model proposed by Davis (1899). The model explains landscape development due to erosion and accumulation processes acting upon structurally controlled bedrock and a river floodplain filled with alluvia, delluvia, and colluvia. *AB* a main longitudinal consequent stream flowing in a trough, *CD, LO, and EF* the lateral consequent streams, *TR and MS* branches of certain lateral consequent streams, *VJQ* lateral ridge, and *GH* headwater

entities and activities, its properties, and its organizational structure (Craver 2006). In the Earth sciences, models are approximations of dynamic landscapes and open systems and can describe multiple processes via different equations (Shen et al. 2023). ***Empirical-statistical modeling*** requires historical data and statistical techniques, which are used to establish a mathematical representation of the system's behavior (Demeritt and Wainwright 2005). ***Statistical*** and ***machine learning models*** require large and robust datasets that are applicable in various situations. However, in many cases, data such as information on the structure and composition of deep Earth mantles, are limited and difficult to obtain. In geomorphology and physical geography, numerical models can be classified into four groups: ***black-box models***, ***process models***, ***mass–energy balance models***, and ***stochastic models*** (Kirkby et al. 1987; Bokulich and Oreskes 2017). In ***black-box models,*** the internal structure of the system is unknown or treated as a unit (Chorley and Kennedy 1971). In the ***machine learning*** domain, black-box models, such as support vector machines, are those that are based on complex mathematical functions, and their outcomes are difficult to explain (Loyola-González 2019). For this reason, several explanatory tools have been proposed in extra packages developed for R (Biecek 2018). Machine learning (ML) and deep learning (DL) methods have been introduced to address a large variety of data, and their primary task is to recognize patterns and regularities in datasets. ML algorithms are often applied to construct a model (mathematical tool) for precise prediction (Kuhn and Johnson 2013). ***Process models*** are based on the empirical relationships between variables, and mass-balance models represent continuous flows of mass and energy between different entities (Bokulich and Oreskes 2017).

Multidimensional datasets consisting of response variables, many predictors, often in different formats (continuous or factor data), and error distributions (e.g., normal, Bernoulli, or bimodal) require a special approach. Some data may require transformation (for example, log or power transformation), and categorical information is sometimes a subject of feature engineering (FE). FE is a process of transforming raw data into formats that are more suitable for modeling, such as ML models (Zheng and Casari 2018). For

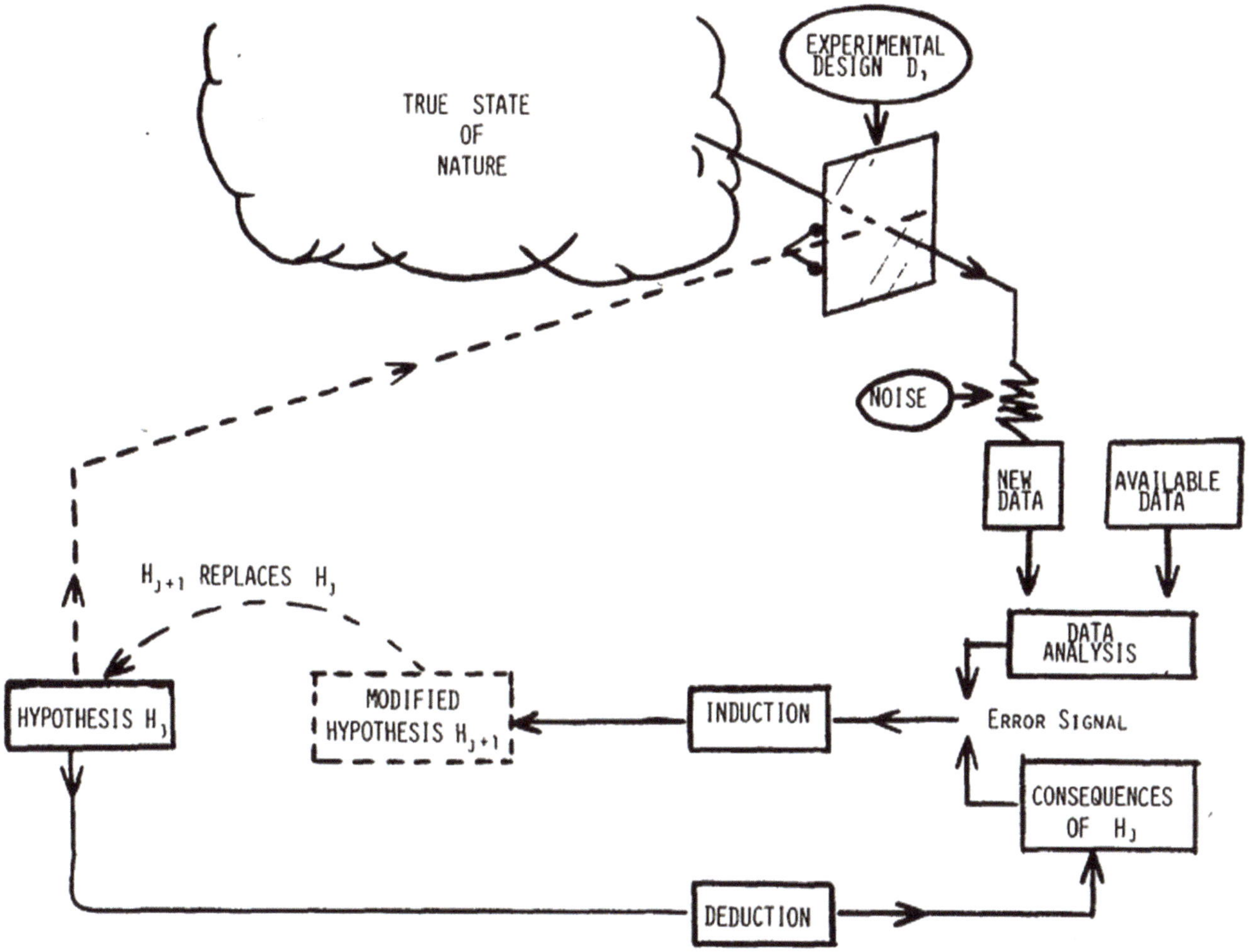

Fig. 1.11 "*Data analysis and data collection in the process of scientific investigation*" (Box 1976). The modeling process is shown as an iterative process. As mentioned by the cited author, "*The experimental design is here shown as a movable window looking into the true state of nature. Its positioning at each stage is motivated by current beliefs, hopes, and fears*" (Box 1976)

example, with individual observations labeled on the basis of tree species names (e.g., beech, oak, and willow), we might want to generalize this information into a single class (deciduous trees), capturing all these species.

Modeling can be a time-consuming process and is frequently considered an iterative procedure based on a heuristic approach (Boehmke and Greenwell 2020). It is based on many important steps, ranging from data cleaning and validation to feature engineering and data transformation to a final model calculation with a selected algorithm. Data contain information that can be simplified to a model form. Owing to the high dimensionality of some datasets, with multiple independent variables (predictors), data cleaning is often the first step, for example, removing highly correlated variables. The simplest numerical model is a linear regression model. In the simplest instance of the model, when only a causal relationship between two variables exists, it can be expressed using a linear model with estimated model coefficients (Fig. 1.12). However, a high correlation does not always indicate real causation.

The most common data analysis methods can be separated into two groups: ***supervised*** and ***unsupervised***. Unsupervised methods do not consider outcome (response or dependent) variables, whereas supervised methods for each observation involve predictor measurement (Kuhn and Johnson 2013; James et al. 2021). For example, supervised methods include linear and logistic regression, whereas an example of an unsupervised method is cluster analysis. In the latter case, we attempt to understand the dataset structure and relationships between variables without considering an unknown response variable. Another group of methods includes ***semisupervised*** modeling techniques. In this approach, a predictive model is constructed by learning from both labeled and unlabeled data. This method is effective for

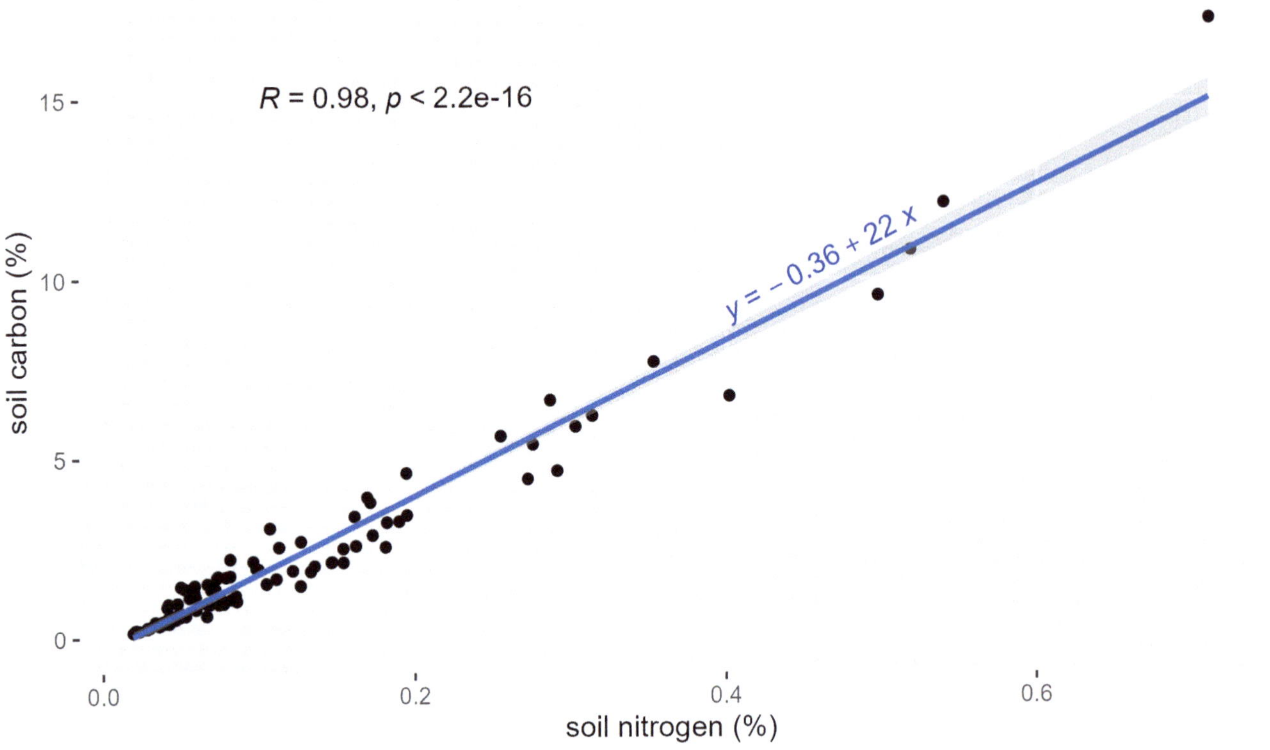

Fig. 1.12 Linear model and its properties. A linear regression model capturing the relationship between soil organic carbon and the nitrogen content in soils on granite bedrock in the Sudety Mountains, Central Europe. Data source: Pawlik et al. (2024)

supervised learning tasks when learning data are scarce or expensive (Fouedjio and Talebi 2022).

Box 1.4 Data in a Nutshell

The data behind Fig. 1.12. Soil carbon and nitrogen contents are closely correlated (Gerrard 1992). Soil is considered the largest terrestrial pool of organic carbon, with an estimated range of 1115–2200 Pg of C (Batjes 2014). High organic carbon and nitrogen contents are typical found in the A horizon and associated with decomposed organic matter and other materials. This relationship is even observed in samples derived from soil profiles beneath living trees. The **ggplot2** package was used to generate simple plots, with some extra functionalities from the **ggpubr** package utilized (Kassambara 2023), including the ***stat_regline_equation()*** and the ***stat_cor()*** functions, which support the input of a model formula, a correlation coefficient (Pearson), and statistical significance (p value). We can see that two variables are highly correlated and that this correlation is significant (is not an outcome of a random process). The relationship is positive; as one variable increases, the other variable also increases. Model performance can be evaluated with the standard ***summary()*** function or with some dedicated packages, such as the **performance** package (Lüdecke et al. 2021). When we set carbon (C) as a response variable and nitrogen (N) as the independent variable, with each unit increase in N, the C content increases by almost 22%. The ***performance::model_performance()*** function provides the following model performance indices: Akaike's information criterion (AIC), AICc, Bayesian information criterion (BIC), goodness of fit (R^2, the proportion of variance explained), adjusted R^2 (the proportion of the variance explained for multiple models), root mean square error (RMSE), and standard deviation (sigma).

```
pacman::p_load("ggpubr")
library(remotes) #a set of functions helping to
download packages from, for instance, GitHub
#Install a package with datasets used in this
book.
```

```
#It can be installed from the author's GitHub
repository.
install_github("Lukpawlik/rappliES")
library(rappliES)
#Get the dataset set from the rappliES package.
data(soils)
library(ggpubr)
#Plot the data.
ggplot(soils, mapping = aes(x=N, y=C))+
 geom_point()+
 geom_smooth(method = "lm")+
 stat_regline_equation(label.x = 0.4, label.y =
 9, color="blue", angle=27)+
 stat_cor(label.x = 0.1, label.y = 15)+
 xlab("soil nitrogen (%)")+
 ylab("soil carbon (%)")
#Saving the plot.
#ggsave("linear_regression.png", height = 10,
width = 17, units = "cm")
library(performance) #assessment of regression
models performance
#We build a model.
cn_model <- lm(C ~ N, data = soils)
#The model can be evaluated using the base R
function summary().
summary(cn_model)
#We can also use the model_performance()
function from the performance package.
model_performance(cn_model)
```

1.8 Prediction, Forecasting, and Prognosis: How Is It Possible At All?

Living and functioning under uncertain conditions or without knowledge of future events is frequently problematic in science, business, and politics. An unpredictable world is unacceptable in modern society but difficult to avoid. It was claimed that "*as a society, we look to predictions to help us make decisions that can mitigate the impact of nature on society and of society on nature*" (Pielke Jr. et al. 1999). ***Prediction*** involves obtaining uncertain information about states and values in an unsampled space or unknown (unmeasured) feature values. A ***prediction*** is also associated with a certain probability that an event will occur in the future. ***Reconstruction*** occurs when we calculate and estimate unknown past events, such as the level of atmospheric CO_2 (Berner 1998). In climatology, ***predictions*** are often presented in the form of probabilities of anomalies in climate variables with lead times up to several seasons (AMS 2012). ***Projection*** is another widely used term, especially when the long-term consequences of climate change are discussed. The Intergovernmental Panel on Climate Change (IPCC) defines the term projection as a potential future evolution of a quantity or set of quantities, often computed with the aid of a model (IPCC 2024). Projections developed by the IPCC include assumptions concerning future socioeconomic and technological developments that may or may not be realized. Predictions of future states on the basis of time series are usually referred to as ***forecasts***. All available information about historical and future events that might impact forecasts is considered (Hyndman and Athanasopoulos 2021). There are important differences between the terms prediction, projection, and forecast, starting from their etymological origin (Silver 2012). The first two terms are of French and Latin origin, and the third originates from Middle English (https://www.etymonline.com/). The word 'prediction' was understood as "*a declaration concerning future events*" ("foretelling"), whereas projection implied "*a throwing forward, a stretching out*". The early meaning of the term "*forecast*" was "*forethought, prudence*", which evolved into a "*conjectured estimate of a future course*". Prediction and forecasting are routinely used interchangeably (https://www.cpc.ncep.noaa.gov/products/outreach/glossary.shtml). Importantly, projection is also a term used in cartography and GISs, meaning the mapping of a curved surface, a sphere or an ellipsoid onto a plane (Lapaine and Frančula 2022).

Some natural events are extremely difficult to predict because of the lack of full knowledge about the complexity of the Earth. However, society and policymakers ask specialists and scientists for help, expecting clear statements about, for instance, geohazards such as tsunamis and earthquakes as potentially devastating events. This expectation arises from the established opinion that if monitoring and scientific activity are financed with public money, society can request trustworthy information about their safety and events that might be hazardous. For earthquakes, a prediction is defined as a future seismic event that may or may not occur in a region at a given time or with a certain magnitude, with a probability p ($0.0 < p < 1.0$) of occurrence (Tapia-Hernández et al. 2019). However, in this case, predictions based on historical data and statistical inference can be misleading, especially when a dataset contains a low number of observations. Therefore, even in cases with low variance, the bias can be high (Fig. 1.13).

As we learned from the L'Aquila earthquake in Italy on 6 April 2009 and from the earthquakes in Turkey and Syria on 6 February 2023, the catastrophic consequences of natural events are not always an outcome of incorrect prediction or the magnitude of the event but rather a matter of communication between different entities involved in risk and public security issues, high concentrations of human populations in active fault zones, or the quality of building structures (Alexander 2014; Zilio and Ampuero 2023). In addition, the quality of a dataset used to train a model can have a pivotal influence on the precision of the predictive model. If the his-

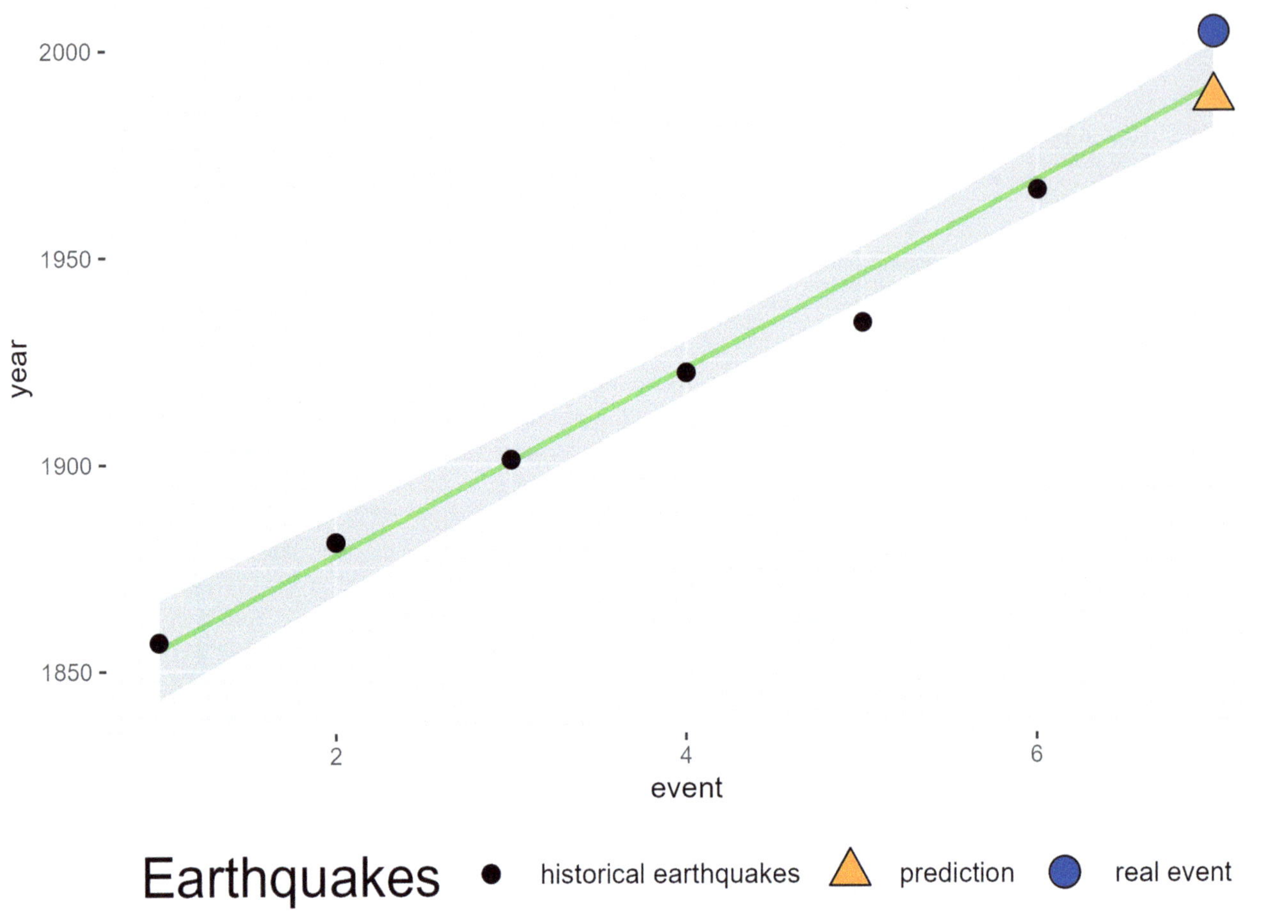

Fig. 1.13 The figure shows how the prediction of earthquake events on the basis of historical data can be uncertain and how a trend develops over time. Data sources: Silver (2012) and Tapia-Hernández et al. (2019). This simplified example shows earthquake events in Parkfield, United States

torical data on debris flows in Norway were collected primarily along roads, the model trained on such a biased dataset could become ineffective when applied to regions with no roads (Lindsay et al. 2022). This rule stems from the concept of the ***area of applicability*** (AOA) in spatial data science (Meyer and Pebesma 2021). The AOA is the area where the model can learn relationships between variables on the basis of the training data and where the estimated cross-validation performance holds (Meyer and Pebesma 2021).

```
earthquakes <- data.frame(year = c(1857, 1881,
1901, 1922, 1934, 1966, 1988, 2004),
 event = c(1, 2, 3, 4, 5, 6, 7, 7))
library(ggplot2)
ggplot()+
  geom_smooth(data = earthquakes, aes(x =
  event, y=year),
              method = "lm", color = "green")+
  geom_point(earthquakes, mapping = aes(x =
  event,
                                     y=year,
          fill = "historical earthquakes",
          size = "historical earthquakes",
          shape = "historical earthquakes"))+
geom_point(earthquakes[7,], mapping = aes(x =
event,
                        y=year,
                        fill = "prediction",
                        size="prediction",
                        shape="prediction"))+
geom_point(earthquakes[8,], mapping = aes(x =
event,
                      y=year,
                      fill = "real event",
                      size = "real event",
                      shape = "real event"))+
scale_fill_manual(name = "Earthquakes",
         breaks = c("historical earthquakes",
         "prediction", "real event"),
         values = c("historical
         earthquakes"="red",
                   "prediction"="orange",
                   "real event"="blue"))+
scale_size_manual(name = "Earthquakes",
```

```
        breaks = c("historical earthquakes",
        "prediction", "real event"),
        values = c("historical earthquakes"=3,
        "prediction"=5,
                                    "real event"=5))+
  scale_shape_manual(name = "Earthquakes",
        breaks = c("historical earthquakes",
        "prediction", "real event"),
        values = c("historical
        earthquakes"=16,
                                   "prediction"=24,
                                   "real event"=21))+
  theme
(legend.position = "bottom",
       legend.title = element_text(size = 20),
       legend.text = element_text(size = 10))
```

References

Achilleos G (2008) Errors within the inverse distance weighted (IDW) interpolation procedure. Geocarto Int 23:429–449

Alexander DE (2014) Communicating earthquake risk to the public: the trial of the "L"Aquila Seven. Nat Hazards 72:1159–1173. https://doi.org/10.1007/s11069-014-1062-2

Altonen B (2012) Valentine Seaman, 1797 (1804)—the black plague or yellow fever of New York City. In: Medical geography and disease surveillance. https://brianaltonenmph.com/gis/historical-disease-maps/valentine-seaman-1804-the-black-plague-or-yellow-fever-in-new-york-city/. Accessed 14 May 2024

AMS (2012) Glossary of meteorology. American Meteorological Society. https://glossary.ametsoc.org/wiki/Welcome. Accessed 10 Feb 2024

Anscombe FJ (1973) Graphs in statistical analysis. Am Stat 27:17–21. https://doi.org/10.1080/00031305.1973.10478966

Aono Y, Kazui K (2008) Phenological data series of cherry tree flowering in Kyoto, Japan, and its application to reconstruction of springtime temperatures since the 9th century. Int J Climatol 28:905–914. https://doi.org/10.1002/joc.1594

Aono Y, Saito S (2010) Clarifying springtime temperature reconstructions of the medieval period by gap-filling the cherry blossom phenological data series at Kyoto, Japan. Int J Biometeorol 54:211–219. https://doi.org/10.1007/s00484-009-0272-x

Arel-Bundock V (2022) Modelsummary: data and model summaries in R. J Stat Softw 103:1–23. https://doi.org/10.18637/jss.v103.i01

ASPRS (2019) LAS Specification 1.4-R15

Baker RE, Peña J-M, Jayamohan J, Jérusalem A (2018) Mechanistic models versus machine learning, a fight worth fighting for the biological community? Biol Lett 14:20170660. https://doi.org/10.1098/rsbl.2017.0660

Batjes NH (2014) Total carbon and nitrogen in the soils of the world. Eur J Soil Sci 65:4–21. https://doi.org/10.1111/ejss.12115

Baynton-Williams R (1969) Investing in maps. Barrie & Rockliff The Cresset Press, London

Berner RA (1998) The carbon cycle and CO2 over Phanerozoic time: the role of land plants. Philos Trans R Soc Lond B 353:75–82

Bickmore DP, Evans IS (1970) Some recent advances in automatic cartography. In: New possibilities and techniques for land use and related surveys with special reference to the developing countries, pp 94–103

Biecek P (2018) DALEX: explainers for complex predictive models in R. J Mach Learn Res 19:1–5

Bingham NH, Fry JM (2010) Regression: linear models in statistics, 1st edn. Springer, London

Bivand RS, Pebesma EJ, Gómez-Rubio V (2008) Applied spatial data analysis with R. Springer

Boehmke B, Greenwell B (2020) Hands-on machine learning with R. CRC Press Taylor & Francis Group

Bokulich A, Oreskes N (2017) Models in geosciences. In: Springer handbook of model-based science. Springer, Cham, pp 891–911

Borkin MA, Vo AA, Bylinskii Z et al (2013) What makes a visualization memorable? IEEE Trans Vis Comput Graph 19:2306–2315. https://doi.org/10.1109/TVCG.2013.234

Box GEP (1976) Science and statistics. J Am Stat Assoc 71:791–799

Box GEP, Jenkins GM, Reinsel GC, Ljung GM (2016) Time series analysis. Forecasting and control, 5th edn. Wiley, Hoboken

Brewer CA (2017) ColorBrewer 2.0. Color advice for cartography. http://www.ColorBrewer.org

Burger SR (2018) Introduction to machine learning with R, 1st edn. O'Reilly Media, Inc., Sebastopol

Burton I (1963) The quantitative revolution and theoretical geography. Can Geogr 7:151–162

Buzai GD (2020) The cholera map by John Snow (London, 1854): a health solution as a conceptual summary of applied geography. An Soc Cient Argent 268:5–18

Cheng J, Schloerke B, Karambelkar B, Xie Y (2023) Leaflet: create interactive web maps with the JavaScript "leaflet" library. R package version 2.2.0

Cheshire J, Kent AJ (2023) Getting to the point? Cartographical J, Rethinking Arrows on Maps. https://doi.org/10.1080/00087041.2023.2178134

Chorley RJ, Kennedy BA (1971) Physical geography. A system approach. Prentice Hall International Inc., London

Clarke KC (2013) What is the world's oldest map? Cartogr J 50:136–143. https://doi.org/10.1179/0008704113Z.00000000079

Craver CF (2006) When mechanistic models explain. Synthese 153:355–376. https://doi.org/10.1007/s11229-006-9097-x

Dacey MF, Marble DF (1965) Some comments on certain technical aspects of geographic information systems. Department of Geography, Northwestern University, Evanston, IL

Davis WM (1899) The geographical cycle. Geogr J 14:481–504

Demeritt D, Wainwright J (2005) Models, modelling, and geography. In: Questioning geography: fundamental debates. Blackwell Publishing, Malden, pp 206–225

Dramsch JS (2020) 70 years of machine learning in geoscience in review. In: Machine learning in geosciences. Elsevier, pp 1–55

Fischer MM, Nijkamp P (1993) Design and use of geographic information systems and spatial models. In: Geographic information systems, spatial modelling and policy evaluation. Springer, Berlin, Heidelberg, pp 1–13

Fotheringham AS, Brunsdon C, Charlton M (2000) Quantitative geography: perspectives on spatial data analysis. Sage Publications Ltd.

Fouedjio F, Talebi H (2022) Geostatistical semi-supervised learning for spatial prediction. Artif Intell Geosci 3:162–178. https://doi.org/10.1016/j.aiig.2022.12.002

Galton A (2019) Space, time and the representation of geographical reality. In: The philosophy of GIS. Springer, Cham, pp 75–97

Gerrard J (1992) Soil geomorphology. In: An integration of pedology and geomorphology. Chapman & Hall, London

Gilbert GK (1880) Report on the geology of the Henry Mountains. Dept. of the Interior, US Geographical and Geological Survey of the Rocky Mountain, Washington, D.C

Godziek J, Pawlik Ł (2023) Indicators of wind-driven forest disturbances—pit-mound topography, its automatic detection and significance. Catena 221:106757. https://doi.org/10.1016/j.catena.2022.106757

Goldsmith B (2005) Obsessive genius: the inner world of Marie Curie. 272 pp, W.W. Norton
Goodchild MF (1992) Geographical information science. Int J Geogr Inf Syst 6:31–45. https://doi.org/10.1080/02693799208901893
Gould P (1979) Geography 1957-1977: the Augean period. Ann Assoc Am Geogr 69:139–151. https://doi.org/10.1111/j.1467-8306.1979.tb01243.x
Guo H (2017) Big earth data: a new frontier in earth and information sciences. Big Earth Data 1:4–20. https://doi.org/10.1080/20964471.2017.1403062
Haggett P (1969) On geographical research in a computer environment. Geogr J 135:497–507
Han J, Kamber M, Pei J (2012) Data mining. Concepts and techniques, 3rd edn. Morgan Kaufmann, Waltham
Hansen MC, Potapov PV, Moore R et al (2013) High resolution global maps of 21st-century forest cover change. Science 342:850–853. https://doi.org/10.1126/science.1244693
Harrell JA, Brown VM (1992) The world's oldest surviving geological map: the 1150 B.C. Turin papyrus from Egypt. J Geol 100:3–18. https://doi.org/10.1086/629568
Hastie T, Tibshirani R, Friedman J (2009) The elements of statistical learning. Data mining, inference, and prediction, 2nd edn. Springer
Heinaro E, Tanhuanpää T, Yrttimaa T, Holopainen M (2021) Airborne laser scanning reveals large tree trunks on forest floor. For Ecol Manag 491:119225. https://doi.org/10.1016/j.foreco.2021.119225
Heywood I, Cornelius S, Carver S (1998) An introduction to geographical information systems. Addison Wesley Longman Limited, New York
Hormann P, Campbell L (2014) Data storage energy efficiency in the zettabyte era. Aust J Telecommun Digit Econ 2:1. https://doi.org/10.7790/ajtde.v2n3.51
Hyndman RJ, Athanasopoulos G (2021) Forecasting: principles and practice, 3rd edn. OTexts, Melbourne
IPCC (2024) The intergovernmental panel on climate change glossary. IPCC. https://apps.ipcc.ch/glossary/. Accessed 10 Feb 2024
James G, Witten D, Hastie T, Tibshirani R (2021) An introduction to statistical learning with applications in R, 2nd edn. Springer
Jenny H (1941) Factors of soil formation. In: A system of quantitative pedology, vol 52. McGraw-Hill, New York, p 415
Kassambara A (2023) ggpubr: "ggplot2" based publication ready plots. R package version 0.6.0
King PB, Schumm SA (1980) The physical geography (geomorphology) of William Morris Davis. Geo Abstracts Limited, Norwich
Kirkby MP, Naden P, Burt T, Butcher D (1987) Computer simulation in physical geography. Wiley, New York
Krumbein WC (1962) The Computer in Geology. Science 136:1087–1092
Kuhn M, Johnson K (2013) Applied predictive modeling. Springer, New York
Lapaine M, Frančula N (2022) Map projections classification. Geographies 2:274–285. https://doi.org/10.3390/geographie2020019
Li J, Heap AD (2014) Spatial interpolation methods applied in the environmental sciences: a review. Environ Model Softw 53:173–189. https://doi.org/10.1016/j.envsoft.2013.12.008
Lindsay E, Frauenfelder R, Rüther D et al (2022) Multi-temporal satellite image composites in Google earth engine for improved landslide visibility: a case study of a glacial landscape. Remote Sens 14:2301. https://doi.org/10.3390/rs14102301
Lisiecki A, Król D (2020) Internet advertising strategy based on information growth in the zettabyte era. In: Advances in computational collective intelligence. Springer Nature, Switzerland, pp 440–452
Lovelace R, Nowosad J, Muenchow J (2019) Geocomputation with R. CRC Press Taylor & Francis Group, Boca Raton, FL
Loyola-González O (2019) Black-Box vs. White-Box: understanding their advantages and weaknesses from a practical point of view. IEEE Access 7:154096. https://doi.org/10.1109/ACCESS.2019.2949286
Lu M, Appel M, Pebesma EJ (2018) Multidimensional arrays for analysing geoscientific data. ISPRS Int J Geo Inf 7:313
Lüdecke D, Ben-Shachar M, Patil I et al (2021) Performance: an R package for assessment, comparison and testing of statistical models. J Open Source Software 6:3139. https://doi.org/10.21105/joss.03139
Mahecha MD, Gans F, Brandt G et al (2020) Earth system data cubes unravel global multivariate dynamics. Earth Syst Dynam 11:201–234. https://doi.org/10.5194/esd-11-201-2020
Meyer H, Pebesma E (2021) Predicting into unknown space? Estimating the area of applicability of spatial prediction models. Method Ecol Evol 12:1620–1633. https://doi.org/10.1111/2041-210X.13650
Midway SR (2020) Principles of effective data visualization. Patterns 1:1–7. https://doi.org/10.1016/j.patter.2020.100141
Miller JR, Turner MG, Smithwick EAH et al (2004) Spatial extrapolation: the science of predicting ecological patterns and processes. Bioscience 54:310–320. https://doi.org/10.1641/0006-3568(2004)054[0310:SETSOP]2.0.CO;2
Montero D, Kraemer G, Anghelea A et al (2023) Data cubes for earth system research: challenges ahead. EarthArXiv. https://doi.org/10.31223/X58M2V
Navaux POA, Lorenzon AF, da Silva SM (2023) Challenges in high-performance computing. J Braz Comput Soc 29:51–62. https://doi.org/10.5753/jbcs.2023.2219
Oreskes N, Shrader-Frechette K, Belitz K (1994) Verification, validation, and confirmation of numerical models in the earth sciences. Science 263:641–646. https://doi.org/10.1126/science.263.5147.64
Pawlik Ł, Gałązka A, Gruba P et al (2024) High-resolution soil sampling reveals the pattern of biological weathering and soil formation under trees. Sci Total Environ 941:173725
Pebesma EJ, Bivand RS (2023) Spatial data science: with applications in R, 1st edn. Chapman and Hall/CRC, Boca Raton, FL
Pielke RA Jr, Sarewitz D, Byerly R Jr, Jamieson D (1999) Prediction in the earth sciences and environmental policy making. EOS Trans AGU 80:309–313. https://doi.org/10.1029/99EO00225
Pigni F, Piccoli G, Watson R (2016) Digital data streams: creating value from the real-time flow of big data. Calif Manag Rev 58:5–25
Pike RJ (2000) Geomorphometry—diversity in quantitative surface analysis. Prog Phys Geogr 24:1–20
Pinsky MA, Karlin S (2011) An introduction to stochastic modeling, 4th edn. Academic, Amsterdam
Playfair W (1786) Commercial and political atlas: representing, by copper-plate charts, the progress of the commerce, revenues, expenditure, and debts of England, during the whole of the eighteenth century. Corry, London
Playfair W (1801) The statistical breviary; shewing, on a principle entirely new, the resources of every state and kingdom in Europe; illustrated with stained copper-plate charts, representing the physical powers of each distinct nation with ease and perspicuity. J. Wallis, London
Reinsel D, Gantz J, Rydning J (2018) The digitization of the world. From edge to core. Framingham: International Data Corporation
Robertson JK (1926) The birth of modern science. Science 23:138–151
Roussel JR, Auty D, Coops NC et al (2020) lidR: an R package for analysis of airborne laser scanning (ALS) data. Remote Sens Environ 251:112061. https://doi.org/10.1016/j.rse.2020.112061
Seaman V (1798) An inquiry into the cause of the prevalence of the yellow fever in New York. T. and J. Swords, New York
Shen C, Appling AP, Gentine P et al (2023) Differentiable modelling to unify machine learning and physical models for geosciences. Nat Rev Earth Environ 4:552–567. https://doi.org/10.1038/s43017-023-00450-9

Shi G (2014) Data mining and knowledge discovery for geoscientists. Elsevier, San Diego

Silver N (2012) The signal and the noise. In: Why so many predictions fail—but some don't. The Penguin Press, New York

Siu-Ngan Lam N (1983) Spatial interpolation methods: a review. The American Cartographer 10:129–149

Skogen MD, Ji R, Akimova A et al (2021) Disclosing the truth: are models better than observations? Mar Ecol Prog Ser 680:7–13. https://doi.org/10.3354/meps13574

Snow J (1855) On the mode of communication of cholera, 1st Ediction, 1849, 2nd edition, 1855. John Churchil, London

Sparks A, Hengl T, Nelson A (2024) GSODR: global summary daily weather data in R

Sui D, Morrill R (2004) Computers and geography: from automated geography to digital earth. In: Geography and technology. Kluwer Academic Publishers

Svensson A, Bigler M, Blunier T et al (2013) Direct linking of Greenland and Antactic ice cores at the Toba eruption (74 ka BP). Clim Past 9:749–766. https://doi.org/10.5194/cp-9-749-2013

Tapia-Hernández E, Reddy EA, Oros-Avilés LJ (2019) Earthquake predictions and scientific forecast: dangers and opportunities for a technical and anthropological perspective. Earth Sci Res J 23:309–315

Tennekes M (2018) tmap: thematic maps in R. J Stat Softw 84:1–39. https://doi.org/10.18637/jss.v084.i06

Tobler WR (1970) A computer movie simulating Urban growth in the Detroit Region. Economic Geography 46:234–240

Tufte ER (2001) The visual display of quantitative information, Second edn. Graphics Press, Cheshire, Connecticut

Tukey JW (1977) Exploratory data analysis. Addison-Wesley, Reading

Turing AM (1950) Computer machinery and intelligence, vol LIX, Mind, pp 433–460

Utrilla P, Mazo C, Sopena MC et al (2009) A palaeolithic map from 13,660 calBP: engraved stone blocks from the late Magdalenian in Abauntz cave (Navarra, Spain). J Hum Evol 57:99–111. https://doi.org/10.1016/j.jhevol.2009.05.005

Vigni ML, Durante C, Cocchi M (2013) Exploratory data analysis. In: Data handling in science and technology. Elsevier, In, pp 55–126

Waters N (2018) GIS: history. In: International encyclopedia of geography: people, the earth, environment and technology. Wiley, New York

Wickham H (2016) ggplot2: elegant graphics for data analysis. Springer, New York

Yano K (2001) GIS and quantitative geography. GeoJournal 52:173–180

Young M, Evans IS (1978) Statistical characterization of altitude matrices by computer. Report 5. Terrain analysis: program documentation. Department of Geography, University of Durham, England

Yu CH (2010) Exploratory data analysis in the context of data mining and resampling. Int J Psychol Res 3:9–22

Zhang K, Chen S-C, Whitman D et al (2003) A progressive morphological filter for removing nonground measurements from airborne LIDAR data. IEEE Trans Geosci Remote Sens 41:872–882. https://doi.org/10.1109/TGRS.2003.810682

Zheng A, Casari A (2018) Feature engineering for machine learning. In: Principles and techniques for data scientists, 1st edn. O'Reilly Media, Inc, Sebastopol

Zhu Y, Zhong N, Xiong Y (2009) Data explosion, data nature and Dataology. In: Brain informatics. Springer, Berlin, Heidelberg, pp 147–158

Zilio LD, Ampuero J-P (2023) Earthquake doublet in Turkey and Syria. Commun Earth & Environ 4:71. https://doi.org/10.1038/s43247-023-00747-z

2 R for Data Science: Functionality and Basic Concepts

Tools, Ürgüp, Türkiye, 2023, photo: Ł. Pawlik

Supplementary Information The online version contains supplementary material available at https://doi.org/10.1007/978-3-031-89673-6_2.

© The Author(s), under exclusive license to Springer Nature Switzerland AG 2025

Ł. Pawlik, *R Applications in Earth Sciences*, Springer Textbooks in Earth Sciences, Geography and Environment,
https://doi.org/10.1007/978-3-031-89673-6_2

> A computer language is described by its syntax and semantics.
> Ihaka and Gentleman (1996)

R is a language and environment for statistical computing and graphics (https://www.r-project.org/about.html (R Core Team 2023)). The R language was developed from the S system invented in 1976 at Bell Laboratories by Rick Becker, John Chambers, and Allan Wilks (https://cran.r-project.org/). R was initially written by Robert Gentleman and Ross Ihaka of the Statistics Department of the University of Auckland in 1991. In 1996, they were finally able to share the message "*(…) we are well on the way to building a complete and useful piece of software*" (Ihaka and Gentleman 1996). The name is partly based on the (first) names of the first two R authors and partly a play on the name of the Bell Labs language "S" (https://cran.r-project.org/) (Fig. 2.1).

Since its first official version (v1.0) released on 29 February 2000, R has gained popularity worldwide and is now widely used in science and industry, supporting various data mining services. This open-source software is available under the terms of the Free Software Foundation's GNU General Public License in source code form. R is now known as the R environment of an ecosystem and consists of basic R software and multiple extensions (libraries and packages) that can be installed as required and increase its functionality. Each package can contain additional functions and sample datasets. R provides several innovations and functionalities that make it highly useful for scientific investigation and data analysis. They are listed on the R project website (https://www.r-project.org/about.html) and include the following:

- An effective data handling and storage facility.
- A suite of operators for array calculations, particularly matrix calculations.
- A large, coherent, integrated collection of intermediate tools for data analysis.
- Graphical facilities for data analysis and display either on-screen or via hardcopy.
- A well-developed, simple, and effective programming language that includes conditionals, loops, user-defined recursive functions and input and output capabilities.

Fig. 2.1 R software official logotype

As with other programming languages (such as Python), the three notable advantages of R are (1) automation of analyses, (2) reproducibility, and (3) extensions in the form of new functions and packages. Because R is a scripting language, all operations can be written in a script and reproduced multiple times with some major or minor modifications, if needed. In addition, with all operations in a script, a user can obtain the required level of precision and understanding for each step during the analysis process. This level of understanding is normally difficult to achieve when software with a graphical user interface (GUI) is used. New custom functions help overcome the limitations of existing packages and the R base package to advance research on new, previously unexplored topics. For example, such expansion has occurred with text analyses (text mining) (Silge and Robinson 2017) and spatial modeling (Pebesma and Bivand 2023). Now, the variety of topics that can be explored in R has surpassed what was once unimaginable, and in this book, we will try to show some approaches related to the Earth sciences.

R can be used as a stand-alone program, but its practical use can be greatly increased with extra packages. As of May 2024, there were as many as 20,705 available packages (Fig. 2.2). These packages only represent those that were available from the Comprehensive R Archive Network (CRAN) repository (https://cran.r-project.org/). However, many others are available from GitHub (https://github.com/) and are under active development. The CRAN is a collection of sites that provide identical material, consisting of R package distributions, contributed extensions, and documentation for R (https://cran.r-project.org/).

```
pacman::p_load("tidyverse", "rvest")
library(tidyverse) #the package installs and
loads multiple 'tidyverse' packages in a single
step
library(rvest) #helps to get information from
web pages
url_packages <- "https://cran.r-project.org/
web/packages/available_packages_by_date.html"
#Get information from CRAN on R packages.
df_packages <- url_packages %>%
  read_html() %>%
  html_nodes("table") %>%
  html_table(fill = TRUE) %>%
```

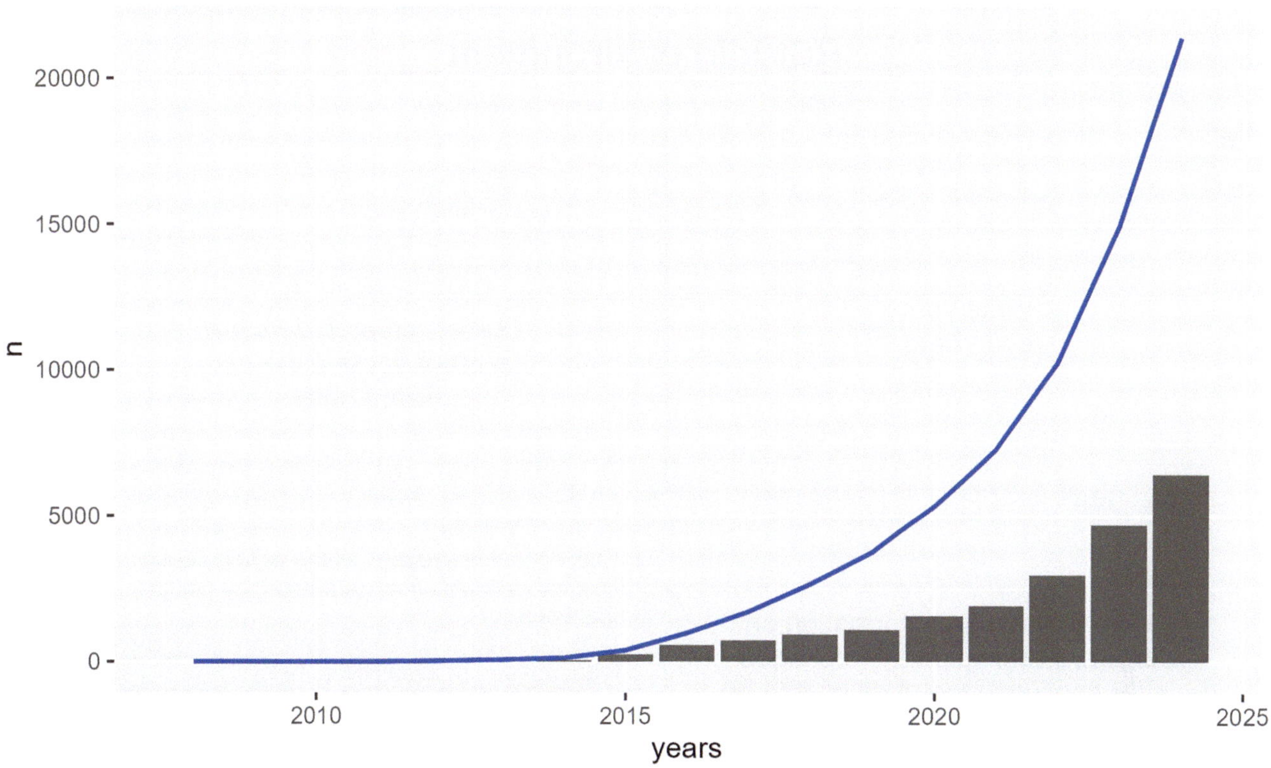

Fig. 2.2 Number of R packages produced annually and available from the CRAN. Data source: https://cran.r-project.org/

```
  as.data.frame() %>%
  mutate(dates = lubridate::ymd(Date)) %>%
  mutate(years = lubridate::year(dates))
#Calculate their number per year.
sum_df_packages <- df_packages %>%
  group_by(years) %>%
  summarize(n = n())
library(ggplot2)
#Plot the data.
ggplot(sum_df_packages)+
  geom_col(mapping = aes(x=years, y=n))+
  geom_line(aes(x=years, y= cumsum(n)), color =
"blue", size = 0.8)
#Save the plot.
#ggsave("Rpackages_number.png", height = 10,
width = 17, units = "cm")
```

When writing scripts, a user can choose from several integrated development environments (IDEs). One of the most popular IDEs for R is RStudio, which consists of a console, a syntax-highlighting editor that supports direct code execution, and tools for plotting, history, debugging, and workspace management (https://posit.co/products/open-source/rstudio/) (Fig. 2.3). The R universe consists of active developers and users who create a large community in which individuals rely on the experience and knowledge of other community members. There is a common opinion that a problem that someone encounters when working with R perhaps has already been solved and that a solution can be easily found via an internet search or by making a request using, for instance, ChatGPT (https://chatgpt.com/). For example, the explanation of a part of a code or a full description of packages, objects, and outcomes can be obtained with ChatGPT (Fig. 2.4). ChatGPT can even be used from inside RStudio using the **chatgpt** package (Rodriguez 2023; Starmer 2023).

Because R is a programming language, it is based on a syntax, which can make first attempting to learn it quite difficult, especially for individuals without prior programming knowledge. Initially, the learning curve can be quite steep. However, multiple resources can be used to overcome early difficulties, allowing user to become quite fluent in R. A good starting point when learning R is accessing free material available online (https://www.r-project.org/doc/bib/R--books.html, https://bookdown.org/).

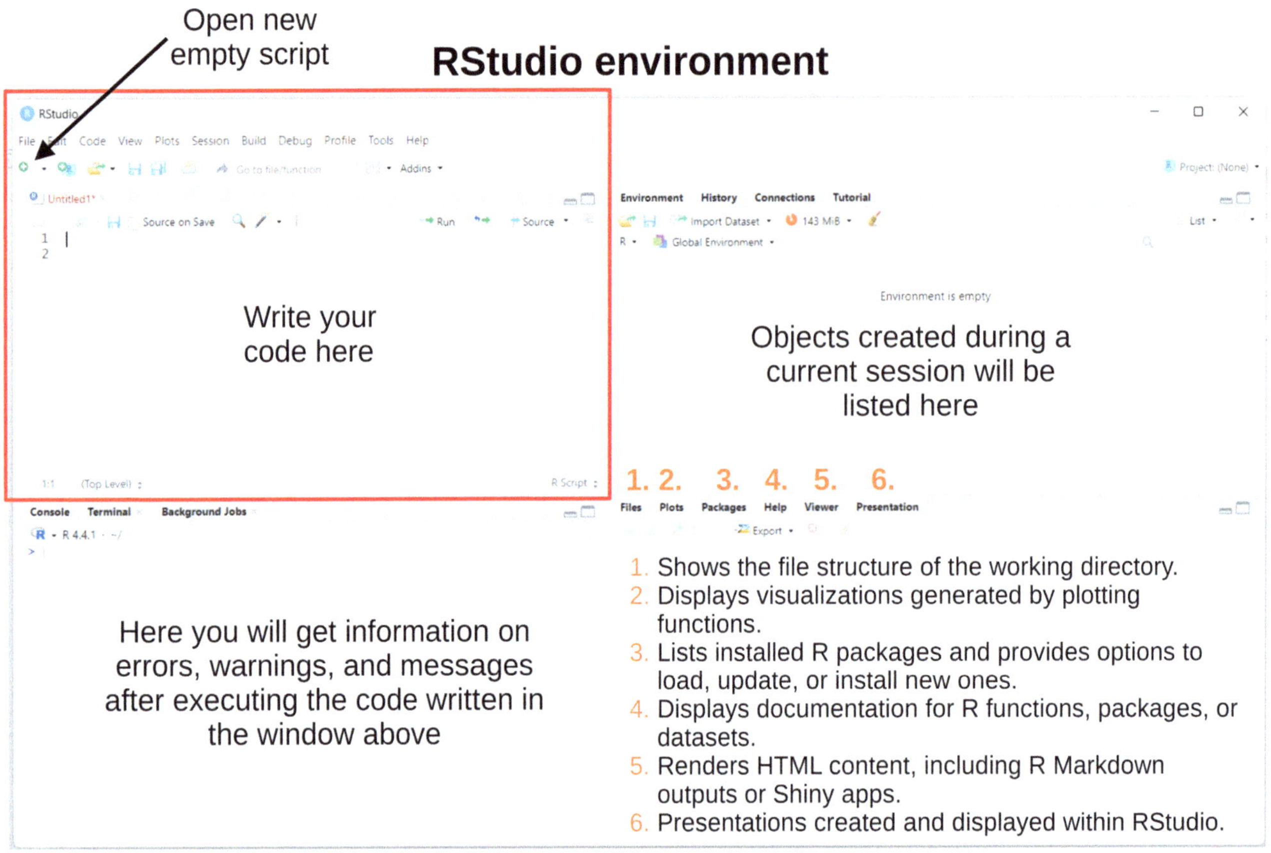

Fig. 2.3 The RStudio main panel with explanations of the four windows where the user writes code (red rectangle), views elements loaded into or created during the current R session (upper right window), receives messages from R after executing code (lower left window), manages plots and packages, and gets help with functions and packages (six panels within the lower right window)

Fig. 2.4 ChatGPT can be used as an effective tool for explaining R code or offering hints to help overcome errors and improve syntax

2.1 R in Practice: Basic Operations

2.1.1 R Objects

R makes use of ***object-oriented*** and ***functional programming***. We recall here the three principles of functional programming proposed by Chambers (2014):

1. Programming consists largely of defining functions.
2. A function definition in a language, such as a function in mathematics, implies that a function call returns a unique value corresponding to each valid set of arguments and only depends on these arguments.
3. A function call has no side effects that could alter other computations.

According to the same author, object-oriented programming is based on four main ideas (Chambers 2014):

1. Everything we compute is an object, and objects should be structured to suit the goals of our computations.
2. The key programming tool is a class definition, which specifies that all objects of a class share the same structure. A structure is defined by properties that are themselves objects of specific classes.
3. A class can inherit from (contain) a simpler superclass such that an object of this class is also an object of the superclass.
4. To perform computations with objects, we can define methods that are only used when the objects are of certain classes.

In R, everything associated with an object (a dynamically created, self-describing container for data) and everything that happens is enabled by a function call (Chambers 2008, 2014). For simplicity, objects represent collections of data of all kinds. Computations in R are based on functions that produce meaningful results, with implementations that can be examined for correctness (Chambers 2008). A ***session*** is a space where everything happens when a user interacts with R. It starts when R is started. When working in R, various objects can be created and used in subsequent analyses and calculations. These objects usually contain some information as a result of an operation or are a data source for further tasks. As an introduction, we first explain the basic types and most commonly used R objects. ***Vectors*** are objects that form the background of most R operations (Peng 2022). This type of object is also called an ***atomic vector*** because it holds data of a single type (character, numeric, integer, logical, or complex). Other frequently used object types are ***lists***, ***data frames***, ***matrices***, and ***time series***. There are various types of spatial objects, such as point and line vectors, polygons and rasters, which are described elsewhere (Chaps. 5 and 6). Lists are collections of vectors that can be of different types. Data frames are collections of vectors that can be of different types but of the same length. They are used to store tabular data and contain rows and columns. Data frames are the most often used objects for statistical analyses. Matrices are regular objects that contain data of an identical type. Additionally, time series contain data with time attributes.

```
#A simple vector of a numeric type.
x <- c(1,2,3) #create an object named x
x #How does it look? Let's call an object and
ask R to show it in the console
#Check the object structure. R automatically
recognizes the variable type.
str(x)
y <- c(1,2,3,"black") #a new vector with
numbers and text
y #How does it look?
str(y) #numbers are converted (coerced, forced)
to characters because it is impossible to
convert a word to a number
#With two vectors containing the same type of
data other operations can be performed.
b <- c(4,5,6)
b
#Creating a new object by adding two vectors.
c <- x+b
c
#Vector can be a sequence of numbers.
x1 <- 1:10 #you can define it in this way
x1
#Every element of a vector can be retrieved by
using square brackets.
x1[3] #what value is on the third position of
vector x?
#How to create a list?
list1 <- list(x, y, b, c, x1) #use the list()
function
str(list1) #what is the structure of the list?
list1
#You can retrieve a selected list element by
using double square brackets.
list1[[1]]
list1[[5]][5] #what is the fifth element of the
fifth vector in the list?
#Now, you can create another object in a matrix
form.
m1 <- matrix(1:8, 4,2) #a matrix containing
numbers from 1 to 8 in two columns and four
rows
m1
str(m1) #what is the structure of the object?
class(m1) #what is the class of the object? Is
it a matrix?
#Can we create it differently?
```

```
d1 <- seq(1, 40, by=1) #create a sequence of
numbers from 1 to 40
d1
#Take the sequence and use it to create a 4x10
matrix.
m2 <- matrix(d1, 4, 10)
m2
d2 <- rep(1:5, 4) #create a sequence of numbers
by repeating it four times
d2
#Take the sequence and use it to create a 5x4
matrix.
m3 <- matrix(d2,5,4)
m3
#Note: you can combine two functions in one
line of code.
m4 <- matrix(seq(1, 40, by=1), 4, 10)
m4
#Another important object type in R is a data
frame (data table).
df1 <- data.frame(col1 = seq(1, 10, by = 1),
#sequence of numbers from 1 to 10
                  col2 = seq(11, 20, by = 1),
#sequence of numbers from 11 to 20
                  col3 = rep("a", 10)) #"a"
                  repeated 10 times
df1
str(df1) #What is the structure of this data
object?
#You can select only the columns you want.
df1[,1] #the first column
df1[,1:2] #the first and the second column
df1[,c(1,3)] #the first and the third column
df1$col1 #or this way
#You can keep the selected columns in a new
object.
df2 <- df1[,c(1,3)]
#Select the first two columns and five top rows.
df1[1:5,1:2]
#Change column names.
colnames(df1) <- c("ColumnA", "ColumnB",
"ColumnC")
df1
```

2.1.2 Data Import, Modification and Transformation

Data form the background and structure of all scientific projects. They frequently originate from external sources and are rarely in a form allowing their direct application without any modification. For this reason, a dataset, before it is used, needs to be thoroughly examined and critically evaluated. There is an important rule of thumb suggesting that analysts should know the data they use (Healy 2018). The garbage in, garbage out (GIGO) concept needs to be mentioned here because it highlights the importance of data quality in modeling and the correct drawing of final conclusions. A low-quality dataset will lead to false or unrealistic results, and data can be messy and require reduction or transformation before use.

In R, a dataset can be a part of a package and serve as a demo dataset, e.g., the **dataset** package (R Core Team 2023). There are also many packages for which the primary task is helping with large-volume data downloads. One example is **GSODR** (Sparks et al. 2024), which is used to download data from the National Oceanic and Atmospheric Administration (NOAA) GSOD web service. Several packages, such as **rsi** (Mahoney 2024), **sits** (Simoes et al. 2021), and **MODISTools** (Hufkens 2023), are used to select and download satellite images and related products. Other packages, such as the **rnaturalearth** package, support the download of various spatial data in raster and vector formats (Massicotte and South 2023). A substantial volume of data can be downloaded directly from the internet using a dedicated URL (uniform resource locator) link and, for example, the ***download.file()*** function. Data can also be imported from a local disc via a file directory and subsequently stored in the temporal memory of the current R session. In most cases, the ***read.csv()*** function is sufficient, but in cases with large datasets, the ***fread()*** function from the **data.table** package can be more convenient (Barrett et al. 2023). The function enables automatic recognition of column separators and classes, the parameters that can sometimes cause problems when the ***read.csv()*** function is used. For Excel spreadsheets, the ***read_excel()*** from the **readxl** package can be used (Wickham and Bryan 2023). Notably, when referring to a specific function, another notation can be used: ***readxl::read_excel()***, which means *"package_name::function_name()"*. In this way, we can save some space and be more precise by adding extra information about the package name.

The most common approach, when working with tabular data, is to use the **dplyr** package (Wickham et al. 2023b), which is a part of the **tidyverse** ecosystem (https://www.tidyverse.org/), which is referred to as an ecosystem because it includes a collection of R packages that share the same philosophy, grammar, and data structures. Wickham et al. (2023a) proposed a model of the data science process that consists of importing elements, wrangling (data cleaning and transformation), visualization, modeling, and communicating. Data cleaning and transformation result in a tabular dataset, where columns are variables and rows are observations. This form allows us to proceed with the subsequent steps. Visualization can assist an analyst at any stage of investigation and provide valuable knowledge of data before the modeling stage. It is a key part of EDA (see Sect. 1.3), but it is also a crucial part of model presentation and explanation.

```
pacman::p_load("dplyr")
#You will now import a file from the Internet.
The file can be from any freely accessible web
service. Let us try to do it by using a known
URL address.
#We start by downloading the data on CO2 and
Greenhouse Gas Emissions. The data set includes
information on CO2 emissions (annual, per
capita, cumulative, and consumption-based),
other greenhouse gases, energy mix, and other
relevant metrics. Before you do that you need
to define a destfile argument in the download.file
function.
#This is a temporary file into which we will
save the data.
temp_co2 <- tempfile()
#Download the file.
download.file("https://nyc3.digitaloceanspaces.
com/owid-public/data/co2/owid-co2-data.csv",
destfile = temp_co2)
#The file has been downloaded and saved in a
temporary directory.
#You can now use it after importing it to the
current session environment as a new object.
co2 <- read.csv(temp_co2)
#Let's inspect the data.
head(co2)#This function will display only the
top six rows of the table. If you want to see
all the data, you can use the View() function
or click on the table icon in the environment
window of the RStudio main panel.
str(co2)#how many observations and variables
does the data set contain?
names(co2)#variable names
summary(co2)#What can be read from the summary
function output?
unique(co2$country)#the data contains
information on 269 territories, including
European Union and non-EU countries
#The data frame will be modified in the
following way:
#1. Reduction - only the following columns will
be selected: country, year, population, gdp,
co2.
#2. The column names will be changed.
co2_mod <- co2[,c(1,2,4,5,8)]
str(co2_mod)
colnames(co2_mod) <- c("Country", "Year",
"Population", "GDP", "CO2")
#You will now do the same by using the dplyr
package. The dplyr package is a part of the
Tidyverse ecosystem (https://www.tidyverse.
org/).
library(dplyr)
#Note the %>% pipe operator. It was introduced
in R around 2013. It allows you to chain
several operations into a step-by-step process.
It makes the code easier to read and follow.
See this post if you want to know more about
the history and meaning of the %>% pipe
operator: https://www.datacamp.com/tutorial/
pipe-r-tutorial
co2_mod1 <- co2 %>%
  select(1,2,4,5,8) %>%
  rename(Country = country, Year = year,
Population = population, GDP = gdp, CO2 = co2)
head(co2_mod1)
tail(co2_mod1)#displaying the last rows (six
rows by default)
#You can follow this operation with the data
transformation. First, the population data will
be divided by 1000; second the GDP will be
log-transformed.
co2_mod2 <- data.frame(co2_mod1[,c(1,2,5)],
Population = co2_mod1$Population/1000,
                     GDP_log = log(co2_mod1$GDP))
head(co2_mod2)
#Note that for some years, there are no values.
They were marked as NAs. How to remove rows
with NA values?
#We can filter the dataset using the
dplyr::filter() function.
co2_mod3 <- co2_mod2 %>%
 dplyr::filter(!is.na(CO2))#filter out NAs values
head(co2_mod3)
#We can get the same results using the subset()
function.
#Note that we use "!=" which means "does not
equal".
co2_mod4 <- subset(co2_mod2, CO2 != "NA")
head(co2_mod4)
```

2.1.3 Tools for Data Visualization in R

The importance of data visualization with R is emphasized in the first sentence of the software description: R is for "*statistical computing and graphics*". Data visualization (DataViz) has been well explained in several R-related books (Chang 2018; Healy 2018; Wilke 2019; Kabacoff 2024). The most powerful and widely used R package dedicated to the construction of visually appealing and meaningful plots is **ggplot2** (Wickham 2016). Its popularity can be partly verified by the number of **ggplot2** extensions available, totaling 133 as of May 2024 (https://exts.ggplot2.tidyverse.org/gallery/). The whole concept of producing graphics with the **ggplot2** package is based on the so-called grammar of graphics (Wilkinson 2005), allowing the definition of each graph component to be made separately as a layer (Wickham 2010). The **patchwork** package is helpful when several ggplot2 plots need to be displayed in a single, combined figure (Pedersen 2024). To do so, mathematical operators are used to combine multiple plots. The **gridExtra** package provides several user-level functions that support multiple grid-based plots on a page and the creation of tables (Auguie 2017). Another well-known R graphical library is the **lattice** package (Sarkar 2008). The visualization method in this package was inspired by William Cleveland's trellis graphics (Cleveland 1994), which is a system that focuses on the display of multivariate data in panels arranged in a regular grid-like structure (Grothendieck 2008). Each panel is used to graph a subset of the data (Becker et al. 1994). Other highly recommended tools are available in the **plotly** package (Sievert 2020). **Plotly** can be used to construct interactive plots that ensure the effective communication of results with the public. This package is powered by the JavaScript library plotly.js. **Highcharter** is another JavaScript-based R library (Kunst 2022). It displays various R objects (data frames, numeric or character vectors, and ts, xts, forecast, or survfit objects) with the ***hchart()*** function. A time series can be elegantly displayed with the **dygraphs** package (Vanderkam et al. 2018). With this package, we can automatically plot *xts* (extensible time series) time series objects (or any object convertible to *xts* format). In addition, this package allows the creation of interactive plots. When a high-quality plot for a publication is needed, the **ggpubr** package is a good starting point (Kassambara 2023). It provides easy-to-use functions for creating and customizing ggplot2-based publication-ready plots. However, to combine ggplot2, lattice, and base R plots in one figure, the **cowplot** package is a better choice (Wilke 2023) (Fig. 2.5). Sometimes, it is better to use a 3-D plot for data display and visualization. Such a task can be accomplished with the **rgl** package (Murdoch and Adler 2024). An important component of data visualization in the Earth sciences is related to data mapping in a scaled geographical space. This is a complex topic that includes thematic mapping (choropleths), dynamic maps (allowing zooming in and out), GIS-based web applications, and digital globes. Maps can be easily constructed with **ggplot2**, but there is also a range of other packages that help handle spatial data and visualize them in the form of beautiful maps (Moraga 2023). **tmap** is an R package that supports the creation of complex maps with all the elements that a proper map should have: a graticule, a legend, a North arrow, and a scale bar (Tennekes 2018). The package is based on the principles of *a layered grammar of graphics,* and the syntax is similar to that of **ggplot2**. The **tmap** package can be used for vector and raster data. Among many functionalities, it can be used to create animated maps (in gif or video formats) with the ***tmap_animation()*** function. Another stage of mapping and displaying the spatial data is reached when the **leaflet** (Cheng et al. 2023) or **mapview** package (Appelhans et al. 2022) is used. Both packages support the construction of interactive maps.

```
#The most popular function used for data
visualization in base R is plot().
plot(co2_mod1$Population, co2_mod1$CO2)
#How can you improve this plot?
#1. You can select data just for one country.
#2. You can change the dot plot into a line
plot.
#3. You can add X-axis and Y-axis labels.
#4. You can add a plot title.
#5. You can change the scientific notation of
the X-axis into real numbers.
czechia <- co2_mod1[co2_mod1$Country ==
"Czechia",]
summary(czechia)
#How to write a similar operation using the
dplyr package?
czechia <- co2_mod1 %>%
 filter(Country == "Czechia")
options(scipen = 999) #This option turns off the
scientific notation.
plot(czechia$Population, czechia$CO2, xlab =
"Population", ylab = "CO2",
     main = "Relationship between the CO2
concentration and population in Czechia")
#You can plot data for 20 territories where CO2
concentrations reached the highest level in
2020.
co2_max20 <- co2_mod1[co2_mod1$Year ==
2020,]#select only 2020
co2_max20 <- co2_max20[order(co2_max20$CO2,
decreasing = TRUE),]#decreasing order
head(co2_max20, 20)
par(xaxt="n") #do not use x-axis labels
#Create custom labels.
lablist_x <- as.vector(co2_max20[1:20,1])
#Plot and save the data. You can save the plot
using the png() function, one of the graphic
devices implemented in R. Note that before
using it, you must define the full directory
path to the folder where the file will be
stored. An example of the path could be "Disc_
partition://My_folder/my_file.png". If you want
to use the png() function, remove the "#"
before it and the dev.off() function. The dev.
off() function shuts down the specified (by
default, the current) device.
#png("co2_popul.png")
plot(co2_max20[1:20,5], type = "h")+
```

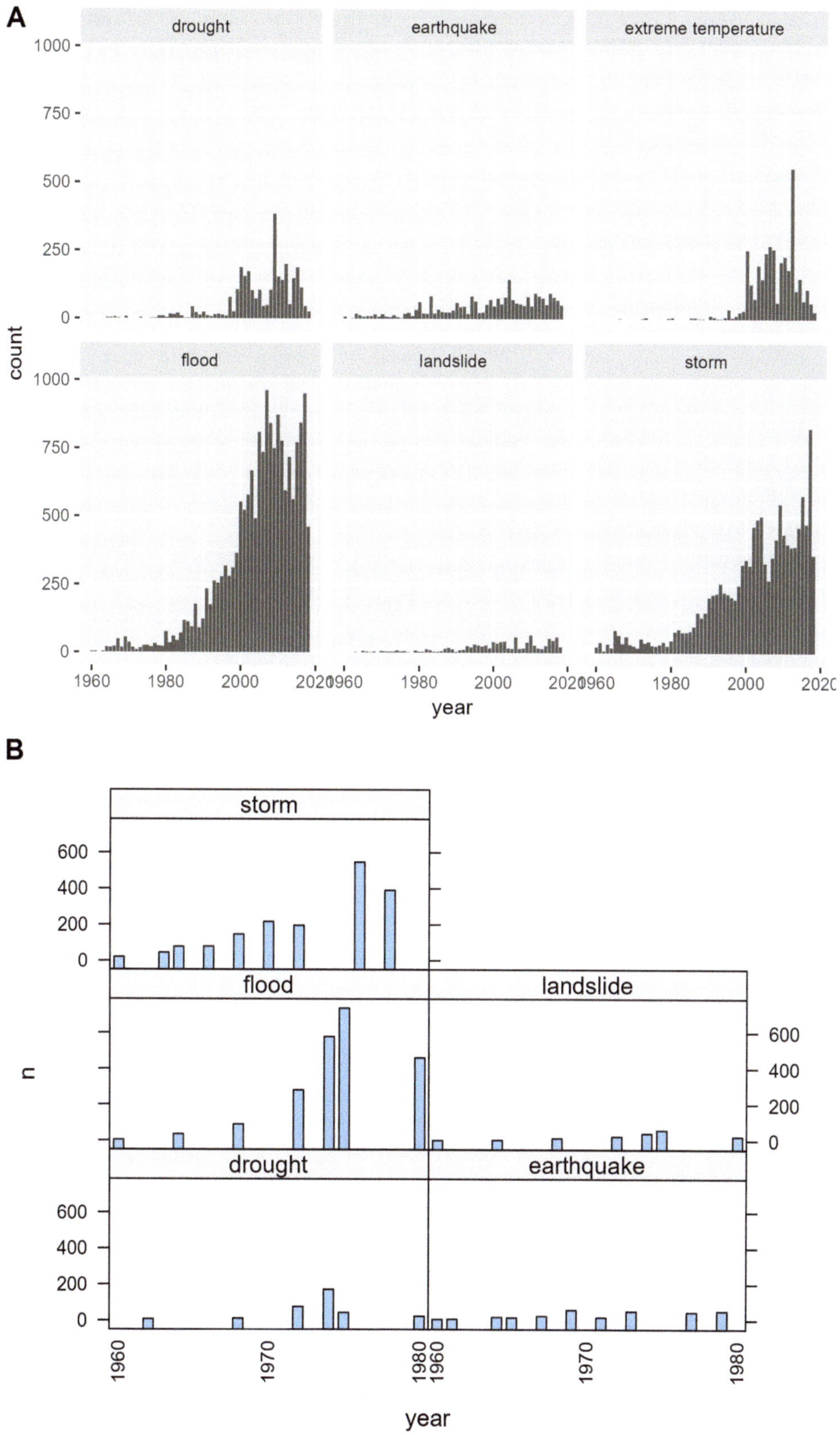

Fig. 2.5 Instances of natural disaster events across the world. An example of ggplot2 (**a**) and lattice (**b**) plots combined into one figure using the **cowplot** package and ***save_plot()*** function

```
axis(1, at=seq(1,20, by=1), labels = FALSE)+
text(x = seq(1,20, by=1), y = 25000, labels =
lablist_x, srt = 90, xpd = TRUE, col = "red")
#dev.off()
#You will now try to do the same exercise by
applying the Tidyverse workflow.
library(ggplot2)
co2_mod1 %>% filter(Year == 2020) %>%
 slice_max(order_by = CO2, n=20) %>%
 ggplot()+
 geom_col(aes(x=reorder(Country, CO2), y=CO2))+
 theme
(axis.text.x = element_text(angle=90))
#Use the ggsave() function to save the plot.
ggsave("co2_popul1.png") #if used without the
name of a specific object, it saves the last
plot created with ggplot2. Note that before
saving the plot, the user has to define the full
path to a local directory. In case of doubts,
call the help panel explanation using a
question mark:
?ggsave()
```

Box 2.1 Data in a Nutshell

Mapping the global and regional distribution of natural disaster events supports mitigating their negative consequences for society and enhancing spatial management and planning. This can be achieved with geocoded data (Resvold and Buhaug 2021). The data used for Figure 2.5 were imported from the Socioeconomic Data and Applications Center (SEDAC), which is a data center that is part of NASA's Earth-Observing System Data and Information System (https://sedac.ciesin.columbia.edu/data/set/pend-gdis-1960-2018/data-download). Unfortunately, the database cannot be downloaded without registering and then logging in to the SEDAC service. After these steps, users gain full access to the data. The data need to be saved in a working directory. The data in this case span the period from 1960 to 2018. This large dataset can be downloaded as a zipped geodatabase (967 Mb) or geopackage (2.1 GB). However, in the present example, we used general information saved and stored in a CSV file (select the file named "*pend-gdis-1960-2018-disasterlocations.csv*"). Plot A was created with the ***geom_bar()*** and ***facet_wrap()*** functions from the **ggplot2** package. Plot B was constructed with the ***barchart()*** function from the **lattice** package. Both plots were combined into one figure with the ***plot_grid()*** function from the **cowplot** package

```
pacman::p_load("dplyr", "ggplot2", "lattice",
"tidyr", "cowplot")
#Before uploading the file with the SEDAC
dataset, we have to decide whether to set the
connection to our working directory. We can set
the connection using the setwd() function. In
this case, when importing and saving files, the
user needs to use only file names and formats
without using the full directory of files.
However, this method is not advisable if
several scripts are open and we do different
operations based on various files and objects.
#Set the connection to your working directory.
setwd("PATH_TO_WORKING_DIRECTORY")
#Download the file manually and save it in your
current session working directory.
#If you use the getwd() function, you will get
information on the current working directory.
getwd()
#Read the file into the session environment. If
the working directory was set, use only the file
name and its extension, as shown below. If the
working directory was not set, use the full
path to the file.
sedac <- read.csv("pend-gdis-1960-2018-
disasterlocations.csv")
library(ggplot2)
library(dplyr)
#Plot the SEDAC data. Note that in this way,
you create a graphical object.
plot1 <- sedac %>%
 filter(!(disastertype == "mass movement (dry)"|
 disastertype == "volcanic activity")) %>%
 ggplot()+
 geom_bar(aes(x=year))+
  facet_wrap(~disastertype)
plot1
#Data summary.
summary(sedac)
#What are the unique types of disasters in this
dataset?
unique(sedac$disastertype)
#We can create a vector with disaster names in
this way:
disaster1 <- c("earthquake", "flood",
"landslide",
             "storm", "volcanic activity",
"mass movement (dry)",
             "drought", "extreme temperature")
disaster1
#Or in this way:
disaster1 <- c(unique(sedac$disastertype))
disaster1
#The 'disaster1' object created during the first
step is replaced by the object created during
```

```
the second step. Two objects cannot have the
same name.
#Count the number of events per disaster type.
sedac1 <- sedac %>%
  group_by(year, disastertype) %>%
  summarise(n = n())
#Or in this way, but here we do not create a
new object.
sedac %>% count(year, disastertype)
library(lattice)#package for data visualization
#Plot the data as a set of scatterplots.
xyplot(n ~ year | disastertype, sedac1)
#Because the plot is difficult to read with each
year label, we create a set of year labels
every decade. Year labels between them are left
empty.
labels <- round(seq(1960, 2018, 10))
labels1 <- c(labels[1], rep(" ", 9), labels[2],
rep(" ", 9),
 labels[3], rep(" ", 9), labels[4], rep(" ",
9),
 labels[5], rep(" ", 9), labels[6], rep(" ",
8))
labels1
#Select some disaster types. However, using the
code in the following line, we get unexpected
results with only 39 observations. For
instance, rows with "extreme temperature"
observations were not selected in this way.
sedac1_filtered <- sedac1[sedac1$disastertype ==
c("earthquake", "flood", "landslide","storm",
"drought", "extreme temperature"),]
sedac1_filtered
#How can we improve it?
#In this way, we get 285 observations.
sedac1_filtered1 <- sedac1[sedac1$disastertype
== "earthquake" |
           sedac1$disastertype == "flood" |
           sedac1$disastertype == "landslide" |
           sedac1$disastertype == "storm" |
           sedac1$disastertype == "drought" |
           sedac1$disastertype == "extreme
           temperature",]
sedac1_filtered1
#We can also use the dplyr::filter() function
for this task.
sedac1_filtered2 <- sedac1 %>%
  filter(disastertype == "earthquake" |
           disastertype == "flood" |
           disastertype == "landslide" |
           disastertype == "storm" |
           disastertype == "drought" |
           disastertype == "extreme
temperature")
sedac1_filtered2
#Plot them using new year labels.
plot2 <- barchart(n ~ year | disastertype,
               data = sedac1_filtered2,
               horizontal = FALSE,
               scales = list(x = list(rot = 90,
                            labels = labels1)))
plot2
#Why is the panel with "extreme temperature"
observations not displayed in this plot?
#Use the unique() function to examine
categories in the disastertype column.
unique(sedac1$disastertype)
#What do you notice?
#It can be difficult to spot, but the category
"extreme temperature"  contains extra space at
the end of the name, resulting in "extreme
temperature ". Please note that in the code
above, we did not use the category name with
the extra space at the end. This caused the
problem because every letter or sign is
interpreted in R. We will repeat the code after
modifying the category name.
sedac1_filtered2 <- sedac1 %>%
  filter(disastertype == "earthquake" |
           disastertype == "flood" |
           disastertype == "landslide" |
           disastertype == "storm" |
           disastertype == "drought" |
           disastertype == "extreme temperature
           ") #note the extra white space
#Plot them using year labels.
plot2 <- barchart(n ~ year | disastertype,
               data = sedac1_filtered2,
               horizontal = FALSE,
               scales = list(x = list(rot = 90,
                            labels = labels1)))
plot2
#The pivot_wider() function is used at this
step, allowing the table to transform from long
to wide. We do that because we want disaster
types as column names. It allows us to
construct separate bar charts for each disaster
type.
library(tidyr) #tools to help create tidy data
where each column is a variable, each row is an
observation, and each cell contains a single
value
sedac_wider <- pivot_wider(sedac1, values_from
= n,
                         names_from = disastertype)
#Change column names for better plotting.
colnames(sedac_wider) <- c("year",
"earthquake", "flood", "landslide",
             "storm", "volcanic_activity",
             "mass_movement",
             "drought", "extreme_temperature")
#Extract column names.
```

```
disaster <- c(names(sedac_wider))
#Plotting.
barchart(earthquake + flood + landslide + storm
+ volcanic_activity + mass_movement + drought +
extreme_temperature ~ year | disaster,
         data = sedac_wider, horizontal =
         FALSE,
         scales = list(x = list(rot = 90,labels
         = labels1)))
#At this step, we combine two plots.
library(cowplot)
final_plot <- cowplot::plot_grid(plot1, plot2,
nrow = 2,
                    labels = c("A", "B"))
final_plot
#Saving directory.
file_dir <- "lattice_ggplot2_polot.png"
#Save the figure.
#save_plot(file_dir, plot = final_plot, base_
height = 10, base_width = 6, base_asp = 1)
```

2.1.4 Export Data and Save the Results of Your Project

Fortunately, analyses performed in R are highly reproducible, and it is easy to perform calculations again if the session is terminated or closed. However, in many instances, we want to save our results in the form of tables, charts, or maps. R has several bitmap file functions that support exporting plots in PNG (good for line drawings or images with solid colors), JPEG (good for photographs or natural scenes), TIFF (used to store raster graphics and image information), and BMP (standard for Windows) formats (Peng 2020). The most basic functions are the ***png()***, ***jpeg()***, and ***tiff()*** function from the **grDevices** package (R Core Team 2023). Plots can also be exported as vector graphics (pdf or svg) and refined via free software (e.g., Inkscape). For this task, the ***pdf()*** and ***svg()*** functions can be used. We can also save the whole R session. In this case, several useful functions can be applied. The ***write.csv()*** function is frequently used for saving data in tabular form. However, with large datasets, the ***fwrite()*** function from the **data.table** package could be a better option. Together with a great variety of graphical packages, there are also different functions for exporting and saving plots and graphics. The ***ggsave()*** function is dedicated to ggplot2 objects, and the ***tmap_save()*** function can be used to save tmap objects. With these functions, several graphic formats are supported, such as PNG, JPG, BMP, or TIFF. When the result of a project is a model, it can be saved and used later for prediction with the ***saveRDS()*** function from the base R suite. The model can be restored under a different name when the ***readRDS()*** function is used.

2.2 Operators in R

In the first part of the book, several operators were introduced, and it is time to explain them in a more systematic way. Undoubtedly, the most important operator is the assignment operator "<-", which assigns a unique name to an object or a value. Other operators include arithmetic, comparison, and logical operators (Table 2.1). When working with code, you will immediately become familiar with another important operator, the dollar sign $. This subsetting operator is used in R to access, add, update and delete elements from named lists, data frames, and other R objects. Another subsetting operator is [], which is used for indexing. Moreover, [[]] is used for element extraction from a list.

```
#Create a data frame object with a list of
operators.
operators <- data.frame(operator = c("+", "-",
"*", "/", "^", "%%", "%/%", "<-", "==", "!=",
">", "<", ">=", "<=", "&", "&&", "|", "||", "!",
":", "%in%", "%*%"),
 name = c("Addition", "Subtraction",
"Multiplication", "Division", "Exponent",
"Modulus", "Integer Division", "Assignment",
"Equal", "Not equal", "Greater than", "Less
than", "Greater than or equal to", "Less than
or equal to", "Element-wise AND", "Logical
AND", "Element-wise OR", "Logical OR", "NOT
```

Table 2.1 List of operators that can be applied in R (Wickham et al. 2023a, b)

Operator	Name	Type
+	Addition	Arithmetic
–	Subtraction	Arithmetic
*	Multiplication	Arithmetic
/	Division	Arithmetic
^	Exponent	Arithmetic
%%	Modulus	Arithmetic
%/%	Integer Division	Arithmetic
<-	Assignment	Assignment
==	Equal	Comparison
!=	Not equal	Comparison
>	Greater than	Comparison
<	Less than	Comparison
>=	Greater than or equal to	Comparison
<=	Less than or equal to	Comparison
&	Element-wise AND	Logical operator
&&	Logical AND	Logical operator
|	Element-wise OR	Logical operator
||	Logical OR	Logical operator
!	NOT operator	Logical operator
:	Sequence of numbers	Miscellaneous
%in%	Element testing	Miscellaneous
%*%	Matrix multiplication	Miscellaneous

```
operator", "Sequence of numbers", "Element
testing", "Matrix multiplication"),
 type = c(rep("Arithmetic", 7), "Assignment",
rep("Comparison", 6), rep("Logical operator",
5), rep("Miscellaneous", 3)))
operators
#Save the table as an image.
library(gt) #create tables from tabular data
operators |>
  gt() |>
  cols_width(everything() ~ px(150)) |>
  gtsave("operators.png")
#Examples.
5+5
5-5
5*5
5/5
5^5
7%%5
7%/%5
6==5
6!=5
6>5
6<5
6>=5
6<=5
#We will now test two logical operators
filtering the data frame with operators. Please
consider two options. The first one with "&"
(AND), and the second one with "|" (OR) logical
operator. What do you get and why?
operators[operators$type == "Arithmetic" &
            operators$type == "Logical
            operator",]
operators[operators$type == "Arithmetic" |
            operators$type == "Logical
            operator",]
#When applying the "&" operator to the type
column, we got zero filtered rows because the
operator type cannot be "Arithmetic" AND
"Logical operator" at the same time. It can
only be one of these types.
#But if we use the "&" operator in this
example, we select one row.
operators[operators$type == "Arithmetic" &
            operators$name == "Addition",]
#NOT operator (!).
#Test if a vector element is not NA.
vector1 <- c(1, 2, 3, NA)
!is.na(vector1)
#Test if 5 belongs to the sequence of numbers
from 1 to 10.
5 %in% 1:10
#Test if "a" or "e" belongs to the set of
letters.
a <- "a"
e <- "e"
a %in% c("a", "b", "c", "d")
e %in% c("a", "b", "c", "d")
#Create matrices.
matrix1 <- matrix(3,3,3)
matrix1
matrix2 <- matrix(5,3,3)
matrix2
#Multiply matrices.
matrix1 %*% matrix2
#Dollar sign $ and its usage.
#To practice the $ operator application to a
named list, first, we convert the operator's
data frame into a list.
operators_list <- as.list(operators)
operators_list
#Extract one list component. Please note the
pop-out window when using the $ sign. It helps
you to select a desired list element.
operators_list$operator
#Add a new list component.
operators_list$new_component <- c(rep("a", 22))
operators_list
#Remove a list component.
operators_list$new_component <- NULL
operators_list
#We will now practice it with a data frame
object. The usage is pretty similar.
#Extract one column.
operators$name
#Add a new column.
operators$new_column <- c(1:22)
#Use the object name or see the Environment
window to see the change.
operators
names(operators)
#Remove a column.
operators$new_column <- NULL
#Use the object name or see the Environment
window to see the change.
operators
names(operators)
```

2.2.1 Pipe Operators

A special group of operators are pipe operators, such as %>% and |>. The %>% pipe operator is from the **magrittr** package (Bache and Wickham 2022). For clarity, here, we refer to a statement from the **magrittr** package developer: *"To achieve its humble aims,* ***magrittr*** *(remember the accent) provides a new "pipe"-like operator, %>%, with which you may pipe a value forward into an expression of function* *call; something along the lines of x %>% f, rather than f(x)"* (Bache and Wickham 2022). With these operators, subsequent code lines can form a logical chain with several levels of operations that are executed one after another. Sometimes, when this convention is used, the code is easier to read and follow than when traditional conventions are used. However, it is also not recommended to write too many code chunks with many, e.g., >10, levels coded with pipe operators. When the **ggplot2** package or its extensions are used, the "+" operator, which combines different **ggplot2** functions, is commonly applied (Fig. 2.6). With the %>% or |> operators, data reading and transformation operations can be streamlined with plotting. This approach can eliminate several lines of code by, for instance, not creating unnecessary objects and, consequently, accelerating your work. The following code chunk shows the striking difference between the base R coding style and a code with pipe operators (Fig. 2.7).

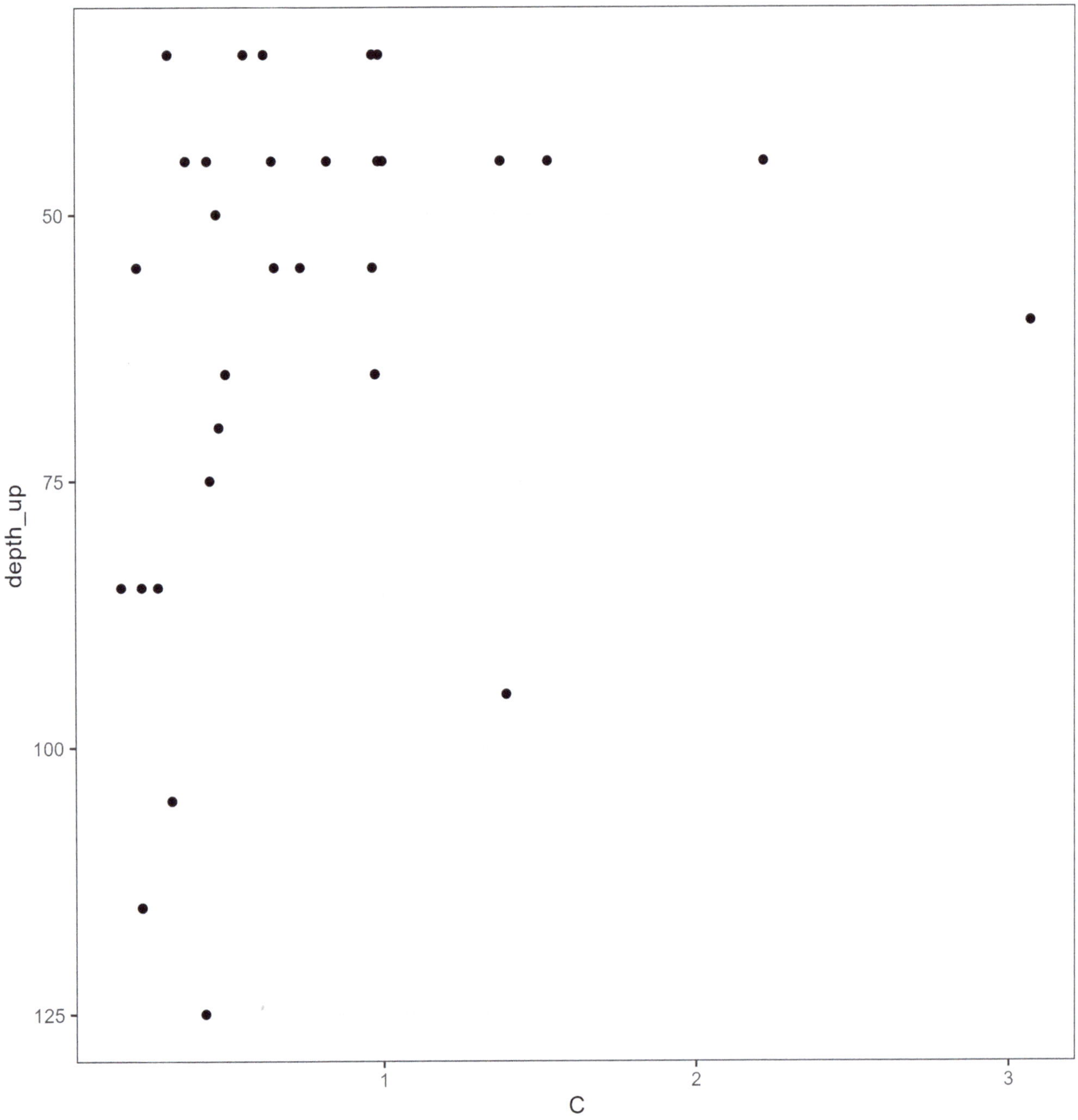

Fig. 2.6 Relationship between the soil organic carbon (OC) content and depth below 30 cm. The figure was constructed with the **ggplot2** package

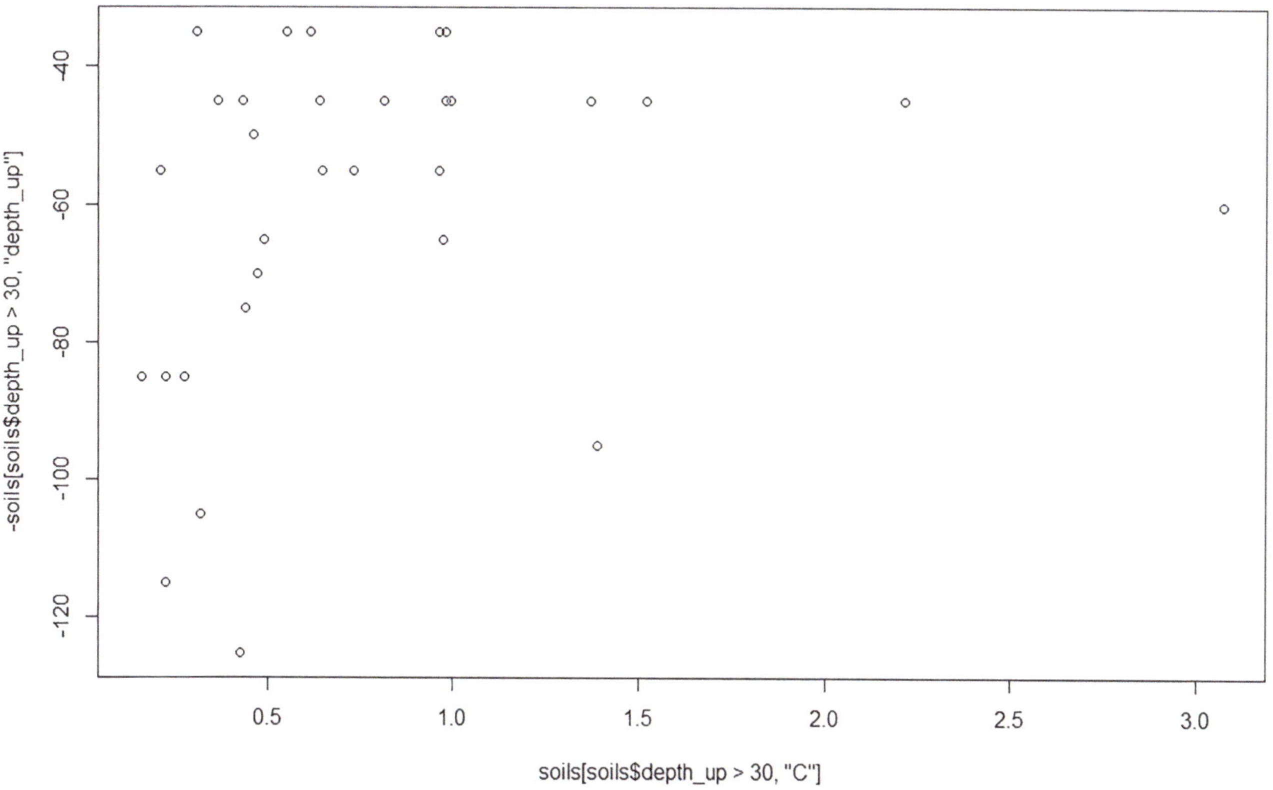

Fig. 2.7 Relationship between the soil organic carbon (OC) content and depth below 30 cm. The figure was constructed with the base R functions

```
#In this example, we use a dataset from the rappliES package. The dataset contains information on several soil properties.
pacman::p_load("rappliES", "ggplot2")
library(rappliES) #custom package with datasets used in this book
data(soils)
#First, we will use |> pipe operator.
soils |>
 dplyr::summarise(meanN = mean(N),
 meanC = mean(C))
#How do you write it using the base R? We write it this way if we do not want to create a new object. Note that we now use the mean() function within the data.frame() function.
data.frame(meanN = mean(soils$N), meanC = mean(soils$C))
#We can also chain data transformation operations using pipe operators, such as filtering and plotting.
library(ggplot2)
soils |>
 dplyr::filter(depth_up > 30) |>
 ggplot(mapping = aes(x=C, y=depth_up))+  # change |> operator to +
 geom_point()+
 scale_y_reverse()+ # reverse the Y axis because it represents the soil depth
 theme_bw()
#ggsave("depth_carbon.png")
#It can be written in this way using the ggplot2 package. We intentionally did not modify the X-axis and Y-axis labels in the two plots created by the following code. How to fix it?
ggplot(soils[soils$depth_up > 30,], mapping = aes(x=C, y=depth_up))+
 geom_point()+
 scale_y_reverse()+
 theme_bw()
#Or using the base R. Note that the "-" operator was added to the second argument to reverse the Y axis. In this example, we select and filter from a data frame using [] (refer to the data frame's rows and columns). By the $ operator, we select the depth_up variable; by the > operator, we filter values over 30.
#png("depth_carbon1.png")
plot(soils[soils$depth_up > 30,"C"], -soils[soils$depth_up > 30,"depth_up"])
#dev.off()
```

2.3 Finding Help After Getting Into Trouble

Owing to the vast community of R developers and users, while working on an R project and getting into trouble, "*you'll never walk alone*" (Rodgers and Hammerstein II 1945). There are several places where you can navigate to and look for extra advice and support. If you are a beginner without prior skills in programming, you will immediately find the R learning curve quite steep. This is entirely normal and should not be a cause of resignation. The price for thousands of functions and solutions offered by R, plus reproducibility, automation, and a full understanding of the R project workflow, is high code clarity and precision. This requires another element to find R extremely useful: your time devoted to learning and practicing. This book is here to help. In every stage of your journey with R, but especially at the beginning, you will find it difficult to produce the desired version of the code without errors and with all functions working smoothly. When programming, handling errors is a daily task. However, as mentioned earlier, there are multiple places and ways to find helpful solutions. The cause of an error can be trivial and easy to find, such as a typo, a missing letter, a file directory that does not exist, an incorrect data format, or NA values in a dataset. First, read the error message you receive in the R or RStudio console. If the explanation is not clear and you struggle with finding an appropriate solution, you can formulate the problem description in your own words or simply copy the error message and paste it in an internet browser or ChatGPT. Alternatively, you can send it to a package developer using, for instance, GitHub. Because R is a programming language and statistical software used worldwide, there is a rule that the user should always follow: it is highly likely that the error you encounter has already been resolved by someone else, and the solution just needs to be found in the global network of the internet. However, it may be that the package you are using is still under active development and testing, and the solution still needs to be formulated. When an error that can be the result of a bug (internal error that causes software malfunction) is encountered, the best solution is to contact package developers, report the error and ask for a solution. In the list below, we summarize the ways in which potential errors in R can be addressed:

1. Read the error message, as it might contain the solution.
2. Carefully read the executed R code that caused the problem. The error could be the result of a typo or other easy-to-find issues.
3. Describe the error or copy and paste it in an internet browser or ChatGPT. StackOverflow (https://stackoverflow.com/) is a popular place where users post questions and obtain solutions from other users and developers.
4. Try to find solutions by reporting errors and asking for assistance from package developers. For example, go to the GitHub site that hosts the **climate** package (https://github.com/bczernecki/climate) for related errors (Czernecki et al. 2020). When you navigate to the Issues section, you will be directed to open and closed issues. The list can be filtered, and if no satisfying solution is found, you can create a new issue. Note that this requires a GitHub user account, which can be obtained for free.

2.4 R Universe and R Knowledge Sources

R is an open-source software that has had a considerable impact on the scientific community and received positive feedback. First, the user community is much larger than that of any commercial software. Second, an expansive community of developers and users means they can each add a small brick to the R ecosystem. Owing to the constant interplay among various members of the community, through websites, blog posts, GitHub profiles, forum discussions, books, conferences, and seminars, the current knowledge that has accumulated since the early days of R is sometimes difficult to filter when searching for appropriate and personalized information. Most likely, the best place to start when looking for extra information and links to other resources is the R Universe portal (https://r-universe.dev/search). Another platform is the R Consortium, which promotes the growth and development of the R language (https://r-consortium.org/). R-bloggers (https://www.r-bloggers.com/) is a collection of posts and articles with R news and tutorials. For a quick start, R cheatsheets are a perfect place to look for information. They are online documents or PDFs that include and explain the most important functions in R, providing a very convenient and fast introduction to package functionality and structure and the logic behind the code. One example is the **lubridate** package (Gudmundsson et al. 2008) cheatsheet https://rstudio.github.io/cheatsheets/lubridate.pdf. This package makes working with dates much easier. Within only two pages, the user can access a large amount of information that is easy to implement and test. Books are another rich source of knowledge, and over 400 are included in a compilation entitled the 'Big Book of R' (https://www.bigbookofr.com/). Most of the books are free of charge and available online. Other ways of learning R include online workshops, lectures, webinars, and short courses, which are easy to find on YouTube. Active participation in international conferences and seminars such as UseR! (https://www.r-project.org/conferences/) or Posit (https://jumpingrivers.github.io/meetingsR/events.html) is the best way of diving into the R community and learning the most recent methods and applications. Countless country-level initiatives and communities that help spread information about R functionalities and applications also exist. One example is the RLadies community and learning platform (https://rladies.org/).

2.5 The Workflow Conventions Used in this Book

This book provides only a snapshot of R's functionality and the ways in which it can support a user's work and projects. Each chapter provides a short introduction to a problem, which is then explored through data and R functions. We start from basic functions, objects, and formats to slowly build confidence when moving through the R ecosystem and various applications for which the software can be used. Many extra tips are provided via the text added to code chunks. Note that the text is introduced following the # sign. Text annotations are added in various places within the code, and for clarity, we use a gray color font to make them different from the R code syntax. This does not affect the code, as it is ignored during execution. Next, the book introduces two approaches to the coding style: base-R coding and the tidyverse approach. We do not argue which approach is better. Both are widely used, and it is good to know and adopt them as needed. For readers interested in digging a bit deeper into a discussion of which workflow is preferred by different users, this post might be useful: https://github.com/matloff/TidyverseSkeptic/tree/master. In many parts of the book, you will see sections of syntax with incomplete file directories, sucha as *#ggsave("file_name.png")*. In these cases, you will need to interact with the code by modifying it, for instance, when reading and writing files. In this case, the code is activated by unchecking it (removing #) and modifying it by adding a full file directory. In most cases, importing or saving files or R objects is necessary. There are three ways to import into and export files from R. The first method is to set a temporary directory to a folder on your machine where you want to save the file or read it from. To use this option, apply the ***setwd()*** function defining the folder directory (for instance, *setwd("H://My_folder/")*). The current directory can be checked with the ***getwd()*** function. With many open R scripts and when working with many files and folders, this method is not recommended. The second method is to set up a new R project in the RStudio console. By using these options, all files will be stored in a folder, which is automatically created when setting up the project. The third way that is used in this book is to define a unique file directory each time we want to import or export a file. For example, if we want to save a **ggplot2** plot, we use the ***ggsave()*** function with just one argument, a file directory consisting of a hard disc partition label, a folder name and a file name with extension (for instance, *ggsave("H://Folder_name/subfolder_name/file_name.png")*). By default, the most recently created **ggplot2** graph or map is saved. A new working folder for saving all plots and maps created with this book is recommended.

Throughout the book, we also work with custom **functions** and **loops** that, in some instances, make calculations and operations easier through their automation and the introduction of new methods and algorithms. Working with functions and loops is not the main objective of this book; for this reason, we recommend looking for other resources that explore these issues much more comprehensively (Gudmundsson et al. 2008; Wickham 2016). We create a new function by defining its name, arguments, and behavior with the following structure: *function_name <- function(x) {function_body}*. A function name should be unique to avoid conflicting with other functions. A function can take several arguments that are used in the function body. Loops are helpful when several identical operations need to be executed. A loop that is defined as *for(variable in sequence) {code to execute}* iterates over a vector or list and executes the same operations for each element of the vector or list. The result of the loop can be stored externally (in a project folder) or in the temporal memory of the current session. When selecting the second option, the user can create an empty list that is then filled with information after each iteration is completed.

Note that in some parts of the book, spatial objects are extensively explored. These objects include points, lines, polygons (known as vectors in GIS software), and rasters (data in grid form). For these specific objects, we normally apply functions from the **sf** package (Pebesma 2018) (for vectors) and **terra** (Hijmans 2023) (for rasters). These objects are more complex than basic objects are because they contain information on the geographical position of each element (points, lines, and raster grid cells) and the attributes of the elements of individual objects; for example, a point can mark the position of a tree, and in its attribute table, information describing the tree species name can be found.

2.6 Working with Packages: Installation and Updating

R is a constantly evolving environment, and for this reason, one should expect that some R libraries will need to be updated (the same with R and RStudio after a new version is released). Additionally, after years of active development, if a package is no longer maintained, i.e., bugs are not fixed, the package may not account for changes that are introduced to other packages. Therefore, various conflicts and incompatibilities can occur among nonupdated and improved packages.

New packages are installed using the ***install.packages()*** function from the **utils** package (R Core Team 2023). Some packages are available from GitHub. For these packages, the ***install_github()*** function from the **remotes** package (Csárdi et al. 2024) is used. When referring to a specific package function, one can use the following notation style: *package_name::function_name()*. Therefore, for the ***install_github()*** function, one can try ***remotes::install_github()***. Throughout the book, we sometimes use this shorthand notation. The packages can be updated in RStudio via the green Update button found in the Packages panel. In this chapter, we include information on all packages and their versions used in this book. This should allow you to determine if you need to install older or newer versions of some packages. The online suplementary materials (scripts) of the book will be constantly updated; please refer to them if problems with package versions are encountered. The first line of each chunk of code text contains a list of packages that must be loaded to have all functions ready to use. For this task, the ***pacman::p_load()*** function is used (Rinker and Kurkiewicz 2017). This is a wrapper for the ***library()*** and ***require()*** functions, and it checks whether a package is installed. If not, it attempts to install it from CRAN or other repositories used by the **pacman** package.

2.6.1 Finding and Installing Older Versions of R Packages

When an older version of the R package needs to be installed, it can be found in the CRAN archive and installed directly using the ***devtools::install_version()*** function (Wickham et al. 2023a, b); for example, *devtools::install_version("dplyr", version = "1.0.7")* will install the **dplyr** package in the 1.0.7 version of R. Another option is to find the package version on CRAN, download the **.tar.gz** file of the package version you need, and install it using the ***install.packages()*** function after defining the full path to the .tar.gz file, for instance, *install.packages ("path_to_package_version.tar.gz", repos = NULL, type = "source")*. Some packages are archived in GitHub repositories. To install them, use *remotes::install_github("github_profile_name/package_name@version")*; for instance, *remotes::install_github("tidyverse/dplyr@v1.0.7")*.

```
pacman::p_load("renv")
#Display information on active packages.
search()
#Get even more information with the
sessionInfo() function. You will get
information on package versions used in the
current session.
sessionInfo()
#Get a list of packages.
loadedNamespaces()
#For effective package and project management,
the renv library is recommended. It helps to
create and manage project-local R libraries.
renv::init()
packages <- renv::dependencies() #find all
package names used in the project
list_of_pacakges <- c(packages$Package)
list_of_pacakges
#To install all packages at once the list can
be used as an argument of the install.
packages() function.
install.packages(list_of_pacakges)
```

References

Appelhans T, Detsch F, Reudenbach C, Woellauer S (2022) Mapview: interactive viewing of spatial data in R

Auguie B (2017) gridExtra: miscellaneous functions for "grid". Graphics

Bache S, Wickham H (2022) magrittr: a forward-pipe operator for R

Barrett T, Dowle M, Srinivasan A (2023) data.table: extention of "data.frame"

Becker RA, Cleveland WS, Shyu M-J, Kaluzny SP (1994) Trellis displays: user's guide. At & T Bell Laboratories, Murray Hill, New Jersey

Chambers JM (2008) Software for data analysis. Programming with R. Springer, New York

Chambers JM (2014) Object-oriented programming, functional programming and R. Stat Sci 29:167–180. https://doi.org/10.1214/13-STS452

Chang W (2018) R graphics cookbook: practical recipes for visualizing data, 2nd edn. O'Reilly Media, Inc.

Cheng J, Schloerke B, Karambelkar B, Xie Y (2023) Leaflet: create interactive web maps with the JavaScript "leaflet" library. R package version 2.2.0

Cleveland WS (1994) The elements of graphing data, 2nd edn. Hobart Press

Csárdi G, Hester J, Wickham H, Chang W, Morgan M, Tenenbaum D (2024) remotes: R Package Installation from Remote Repositories, Including 'GitHub'. R package version 2.5.0, https://CRAN.R-project.org/package=remotes

Czernecki B, Głogowski A, Nowosad J (2020) Climate: an R package to access free in-situ meteorological and hydrological datasets for environmental assessment. Sustain For 12:394. https://doi.org/10.3390/su12010394

Grothendieck G (2008) Lattice: multivariate data visualization with R. J Stat Softw 25:1–3. https://doi.org/10.18637/jss.v025.b02

Gudmundsson MT, Larsen G, Höskuldsson Á, Gylfason ÁG (2008) Volcanic hazards in Iceland. Jökull 58:251–268

Healy K (2018) Data visualization. A Practical Introduction. Princeton University Press

Hijmans R (2023) terra: spatial data analysis. R Package Version 1(7–1):8

Hufkens K (2023) The MODISTools package: an interface to the MODIS land products subsets web services

Ihaka R, Gentleman R (1996) R: a language for data analysis and graphics. J Comput Graph Stat 5:299–314

Kabacoff R (2024) Modern data visualization with R. Chapman & Hall

Kassambara A (2023) ggpubr: "ggplot2" based publication ready plots. R package version 0.6.0

Kunst J (2022) highcharter: a wrapper for the "Highcharts" library

Mahoney M (2024) rsi: efficiently retrieve and process satellite imagery

Massicotte P, South A (2023) rnaturalearth: world map data from natural earth

Moraga P (2023) Spatial statistics for data science. Theory and Practice with R. Chapman and Hall/CRC

Murdoch D, Adler D (2024) rgl: 3D visualization using OpenGL

Pebesma E (2018) Simple features for R: standardized support for spatial vector data. R J 10:439–446. https://doi.org/10.32614/RJ-2018-009

Pebesma EJ, Bivand RS (2023) Spatial data science: with applications in R, 1st edn. Chapman and Hall/CRC, Boca Raton, FL

Pedersen TL (2024) Patchwork: the composer of plots

Peng RD (2020) Exploratory data analysis with R. Lulu. com

Peng RD (2022) R programming for data science

R Core Team (2023) R: a language and environment for statistical computing

Resvold EL, Buhaug H (2021) GDIS, a global dataset of geocoded disaster locations. Sci Data 8:61. https://doi.org/10.1038/s41597-021-00846-6

Rinker TW, Kurkiewicz D (2017) Pacman: package management for R

Rodgers R, Hammerstein II O (1945) You'll Never Walk Alone (Song), Carousel (Musical).

Rodriguez JC (2023) chatgpt: Interface to ChatGPT from R. R package version 0.2.3. https://CRAN.R-project.org/package=chatgpt

Sarkar D (2008) Lattice: multivariate data visualization with R

Sievert C (2020) Interactive web-based data visualization with R, plotly, and shiny. Chapman and Hall/CRC

Silge J, Robinson D (2017) Text mining with R. O'Reilly Media, Inc

Simoes R, Camara G, Souza F et al (2021) Satellite image time series analysis for big earth observation data. Remote Sens 13:2428. https://doi.org/10.3390/rs13132428

Sparks A, Hengl T, Nelson A (2024) GSODR: global summary daily weather data in R

Starmer J (2023) chatgpt: Interface to OpenAI's ChatGPT API. R package version 0.2.0. https://CRAN.Rproject.org/package=chatgpt

Tennekes M (2018) tmap: thematic maps in R. J Stat Softw 84:1–39. https://doi.org/10.18637/jss.v084.i06

Vanderkam D, Allaire J, Owen J, et al (2018) dygraphs: Interface to "Dygraphs" interactive time series charting library

Wickham H (2010) A layered grammar of graphics. J Comput Graph Stat 19:3–28. https://doi.org/10.1198/jcgs.2009.07098

Wickham H (2016) ggplot2: elegant graphics for data analysis. Springer, New York

Wickham H, Bryan J (2023) READXL: Read excel files

Wickham H, Çetinkaya-Rundel M, Grolemund G (2023a) R for data science. Import, tidy, transform, visualize, and model data, 2nd edn. O'Reilly Media, Inc

Wickham H, François R, Henry L, et al (2023b) DPLYR: a grammar of data manipulation

Wilke CO (2019) Fundamentals of data visualization. A primer on making informative and compelling figures, 1st edn. O'Reilly Media, Inc

Wilke C (2023) Complot: streamlined plot theme and plot annotations for "ggplot2"

Wilkinson L (2005) The grammar of graphics, 2nd edn. Springer

3 Soil Data Structure, Properties, and Visualization

Soil profile, Hà Giang, Xuan Minh, Vietnam, 2022, photo: Ł. Pawlik

Supplementary Information The online version contains supplementary material available at https://doi.org/10.1007/978-3-031-89673-6_3.

© The Author(s), under exclusive license to Springer Nature Switzerland AG 2025

Ł. Pawlik, *R Applications in Earth Sciences*, Springer Textbooks in Earth Sciences, Geography and Environment,
https://doi.org/10.1007/978-3-031-89673-6_3

Each soil has had its own history.
Charles Kellogg (1941)

Soil feature data are important sources of field-based information when the potential fertility of a site, such as for food production, requires thorough evaluation. The pedosphere, consisting of soil layers, units, and pedons, is often referred to as the Earth's living skin (epiderm). The pedosphere envelopes the land surface and sustains life on our planet. Therefore, measuring and monitoring soil properties and mapping soil units for spatial planning and management are crucial tasks. Soils are four-dimensional systems with components that constantly change and evolve (Schaetzl and Thompson 2015). They can, however, be degraded due to mass movements (landslides), water and aeolian (wind) erosion, or human impacts. The International Union of Soil Sciences (IUSS) defines material that can be considered soil as "*any material within 2 m of the Earth's surface that is in contact with the atmosphere, excluding living organisms, areas with continuous ice not covered by other material, and water bodies deeper than 2 m*" (IUSS Working Group WRB 2022). There is a long list of soil features that can be measured and described, such as soil structure, depth, texture, color, consistency, and porosity (https://www.fao.org/soils-portal/data-hub/soil-properties/physical-properties/en/). All of these properties require extensive fieldwork and laboratory analyses to be fully understood. Because several steps are required to obtain full knowledge of soils, starting from field sampling and description, there is a certain fraction of uncertainty associated with each observation. In addition, soil collection, analysis, and interpretation are expensive and require a substantial financial budget, especially when several analyses are to be completed for a large area. Finally, some places are extremely difficult to reach. Therefore, we may never obtain comprehensive knowledge of soils or complete coverage of the land surface with detailed soil information. Owing to these factors, scientists use indirect data to estimate potential soil properties.

Notably, soil data are a type of discrete information that is normally collected along one soil profile at a certain depth (1–2 m, if possible). In some cases, soil samples are collected from a pit or trench following a systematic sampling pattern (procedure) along a vertical profile. The soil material can also be collected using a handheld soil auger. The soil sampling position may follow the soil horizons, i.e., naturally formed layers most often oriented horizontally, or a fixed sampling interval (10–20 cm) can be adopted when the soil horizons are uncertain. The first stage of data collection has a considerable effect on the quality of the soil data obtained during further laboratory analyses. The quality of data can be poor if, for example, samples are contaminated, the material is unintentionally mislabeled, or one sample contains material from other samples. Data interpretation allows the formulation of conclusions for 'soil units', and with the results for a complete set of soil units in an investigated area, a soil map can be constructed. Uncertainty is an inherent property of methods used for soil sample collection and physical and chemical evaluations of soils. The lack of complete information in soil studies arises from the nature of the soil itself. Soil is a continuous layer that is sampled only at specific points that are assumed contain the best information needed to correctly classify soil units. The general lack of systematic information can be offset with data obtained from indirect sources, such as via spatial interpolation. The result of this process is a transition from point (discrete) to raster (continuous) data. This transformation is a key step in digital soil mapping.

3.1 Soil Data Structure, Properties, and Visualization

The most intuitive soil data structure is tabular when rows follow soil horizons. The main horizons are normally easily recognizable during direct inspection; for example, the A horizon is a topsoil horizon with a high amount of organic matter. However, there are many instances of homogeneous soils, such as those formed from deeply weathered bedrock, when horizons are not easily recognizable. Additionally, some immature soils that are thin or contain large volumes of coarse rock fragments, e.g., Leptosols or Regosols (the nomenclature follows the World Reference Book), can be difficult to discern. Variables indicating the physical and chemical features of the soil are displayed in columns. When several profiles are condensed in one dataset, the samples need to be labeled accordingly. The variables characterizing soil properties can be of different types (e.g., numeric or categorical). Additionally, numerical variables are usually expressed in different units. Soils can be characterized by their physical and chemical properties. The most widely used soil physical feature is the particle size distribution, which is based on the most general division of soil particles into sand, silt, and clay particles (2–0.063 mm, 0.063–0.002 mm, and <0.002 mm in diameter, respectively). This separation of soil fraction classes is performed according to the World Reference Base for Soil Resources of the International Union of Soil Sciences (IUSS Working Group WRB 2022), which is one of the two most widely used classification systems (Corral-Pazos-de-Provens et al. 2022). Information about the physical properties of the soil can be nicely summarized with ternary plots (Fig. 3.1). This knowledge is crucial for the appropriate evaluation of soil fertility and susceptibility to erosion and directly influences other soil properties, such as bulk density and water content. For example, sandy soils do not effectively retain water and nutrients that are washed from highly permeable soil bodies. Soils with a high proportion of clay can be very

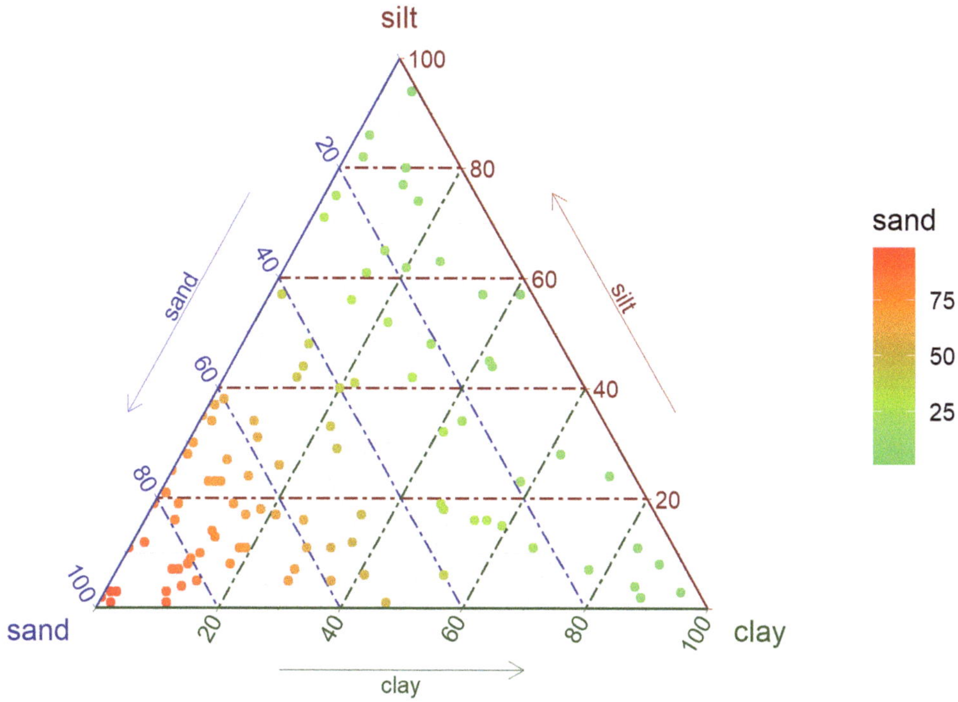

Fig. 3.1 Ternary diagram. A ternary plot constructed with the **ggtern** package (Hamilton and Ferry 2018). The example is based on randomly generated data

soft when wet but very hard when dry. These conditions can limit the growth of some plants. The soil pH and organic carbon (C_{org}) and nitrogen (N) contents are among the most commonly measured chemical soil properties. The soil pH is a quantitative indicator of the acidity and alkalinity of a soil solution and ranges from 0 to 14 (Schaetzl and Thompson 2015). pH 7 is neutral, <7 is acidic, and >7 is alkaline. Acidity or alkalinity is not a linear metric but is instead logarithmic. Soil organic carbon contains atmospheric CO_2 transferred to soil organic matter (SOM) and stabilized in the soil, typically in the upper horizons, but is also stored deeper as dissolved organic carbon (DOC) (Dubeux et al. 2024). Nitrogen is added to the soil naturally through biotic processes associated with plant decomposition, N_2 fixation by the soil biota, or rainfall (Post et al. 1985).

To illustrate the application of a ternary plot, we use randomly generated data with sand, silt, and clay contents.

```
pacman::p_load("ggtern")
sand_num <- sample(1:98, 100, replace = TRUE)
silt_num <- 100-sand_num
function1 <- function(x){
 x1 <- 100-x
 sample1 <- sample(1:x1, 1)
 return(sample1)}
silt1 <- purrr::map_vec(sand_num, function1)
df_soil1 <- data.frame(sand = sand_num,
 silt = silt1,
 clay = 100-(sand_num+silt1))
library(ggtern) #supports ternary diagrams plotting
ggtern(data = df_soil1, aes(x=sand, y=silt, z=clay,
color=sand))+
 geom_point()+
  scale_color_gradient(low="green", high = "red")+
  theme_rgbw()
#Save the file.
ggtern::ggsave("plot.png")
```

Box 3.1 Data in a Nutshell

This example is based on randomly generated data produced with the ***sample()*** function and a simple custom function that allows random selection and calculation of the percentages of the sand, silt, and clay fractions. In theory, for each analyzed soil sample, these values should sum to 100%. For such data, a ternary (triangle) diagram can be used. This plotting approach was first proposed by German astronomer, mathematician, and cartographer Tobias Mayer around 1746–1752 (Howarth 1996). Mayer's color triangle is a palette in which the primary colors are red lead, yellow ochre, and Berlin blue. This method of data display is sometimes known as Gibb's triangular diagram (Pohlmann 2020) or a Ferret ternary plot (Yongue-

Fouateu et al. 2016). Figure 3.1 was prepared with the **ggtern** package (Hamilton and Ferry 2018), which is an extension of the **ggplot2** package (Wickham 2016). Each point represents three values. The ternary diagram should be read in this way: a line parallel to the silt axis (brown) and passing through a selected point on the sand axis is considered. In this case, the value from the sand axis (blue) is read. A color scale is used to display information regarding the soil fraction. The plot could be more complex, but as with other types of data, we need to balance the amount of data shown in one plot and its readability. Other R libraries designed for ternary plot preparation include, for instance, the **Ternary** package (Smith 2017).

3.2 R in Practice: Soil Data Analysis and Visualization

3.2.1 Data in Context

Here, we focus on a set of data from soil profiles sampled in Gorce National Park (https://gpn.gov.pl/), southern Poland. The data were part of a larger project focused on soil development and properties in areas below trees growing on different geological substrates (Pawlik and Samonil 2018). The data were collected from soil profiles excavated under silver fir (*Abies alba*) and European beech (*Fagus sylvatica*) trees (Fig. 3.2). The main aim of this study was to evaluate the properties of soils under the influence of different trees. It was assumed that soils that develop under trees have different physical and chemical properties because of differences in the concentration of water, microbial activity, root system expansion, and carbon content (Pawlik et al. 2023, 2024). The different soil-forming conditions under trees were previously estimated using indirect methods, including electrical resistivity tomography (ERT), and reported (Pawlik and Kasprzak 2018). The focus of this study was biological weathering and biotic–abiotic interactions at the interface between plants and bedrock.

3.2.2 Soil Data Analysis and Visualization

A collection of R packages dedicated to soil analysis and related tasks can be found on the R project website under the task views subpage (https://cran.r-project.org/). The **aqp** R package (Beaudette et al. 2022)—Algorithms for Quantitative Pedology—provides a general toolkit for soil scientists and is perfect for the introduction of soil science concepts. The full tutorial can be found here: https://ncss-tech.github.io/AQP/. However, the first steps allowing us to gain some initial knowledge of the data at hand can be performed via more general tools applied during EDA. This includes the calculation of descriptive statistics, assessing the completeness of the database (*are there any missing observations?*), determination of the variable distribution, examination of the density plots, and analyses of the correlations between variables. Several packages support simple summaries of data, one of which is the **modelsummary** package (Arel-Bundock 2022). With its tools, users can create tables and plots to obtain a general summary of the data and statistical models in R (Tables 3.1 and 3.2). A basic table with descriptive statistics can be built with the ***datasummary()*** function (Table 3.1). Another function, ***datasummary_skim()***, automatically achieves the same goal, but when the output is saved to a PNG file in some cases, histograms are not visible. Other R packages that support summary table preparation include, for instance, the **vtable** package (Huntington-Klein 2023) (Table 3.2).

```
pacman::p_load("rappliES", "modelsummary",
"vtable")
library(rappliES) #custom library with datasets
used in this book
#For this task, we will use soil data from
Gorce National Park, S Poland
data(gorce_soils)
#str(gorce_soils)#the table contains 60
observations of 30 variables
library(modelsummary) #create summary tables
datasummary(All(gorce_soils[,2:11]) ~ Mean + SD
+ Min + Median + Max + Histogram,
 data = gorce_soils)
#The export option in the RStudio Viewer window
can be used to save the table.
#Other options for data summary include, for
instance, the vtable package.
library(vtable) #create data documentation
st(gorce_soils[,2:11])
```

After obtaining information on the basic descriptive statistics, the next step is to calculate and visualize a correlation matrix. Correlation is a measure of a monotonic association between two variables (Schober et al. 2018). This task can be achieved with the help of several tools implemented in various R packages, such as by using the ***chart.Correlation()*** function from the **PerformanceAnalytics** package (Peterson and Carl 2020). With this function, we can define which method should be used to calculate the correlation coefficient: the Pearson (Pearson's product moment correlation, r), Kendall (Kendall's tau, T), or Spearman (Spearman's rank order correlation, ρ, rho) methods. Because all variables in the Gorce soils database are continuous and more or less follow a normal distribution (which can be tested through the Shapiro–

Fig. 3.2 Soil profile under a fir tree stump in Gorce National Park (all photos by Pavel Šamonil, reproduced with the permission of the author). (**a**) General view, (**b**) control profile, (**c**) taproot growing between two sandstone blocks, (**d**) general view of the root system architecture, and (**e**) decomposing root and empty root channel. Description of images after Pawlik and Samonil (2018)

Table 3.1 Descriptive statistics computed for selected properties of soil samples collected in Gorce National Park

	Mean	SD	Min	Median	Max	Histogram
Top	25.17	23.54	0.00	20.00	90.00	
Bottom	35.42	28.09	5.00	30.00	110.00	
pH_H$_2$O	4.60	0.25	4.14	4.60	5.17	
pH_KCl	3.74	0.17	3.34	3.72	4.09	
Cox	1.39	1.16	0.33	1.11	7.49	
N	0.10	0.08	0.02	0.08	0.51	
Q4_6	12.08	7.64	0.52	12.47	34.90	
C_THS	1.00	0.66	0.16	0.85	2.86	
C_HA	0.34	0.20	0.08	0.27	1.33	
C_FA	0.66	0.50	0.03	0.56	2.07	

The table is based on data from Pawlik and Samonil (2018)

top depth of the topsoil horizon (in cm), *bottom* depth of the bottom soil horizon (in cm), *pH_H2O* soil pH determined in water, *pH_KCl* soil pH determined in KCl solution, *Cox* oxidizable carbon, *N* total nitrogen content, *Q4_6* color quotient, *C_THS* carbon content in total humic substances, *C_HA* carbon content in humic acids, and *C_FA* carbon content in fulvic acids

Table 3.2 Descriptive statistics for the Gorce soil data calculated using the *vtable::st()* function (Huntington-Klein 2023)

Variable	N	Mean	Std. Dev.	Min	Pctl. 25	Pctl. 75	Max
Summary statistics							
Top	60	25	24	0	5	40	90
Bottom	60	35	28	5	10	50	110
pH_H_2O	60	4.6	0.25	4.1	4.5	4.8	5.2
pH_KCl	60	3.7	0.17	3.3	3.6	3.9	4.1
Cox	60	1.4	1.2	0.33	0.63	1.9	7.5
N	60	0.1	0.082	0.023	0.05	0.13	0.51
Q4_6	60	12	7.6	0.52	7	17	35
C_THS	60	1	0.66	0.16	0.55	1.2	2.9
C_HA	60	0.34	0.2	0.079	0.25	0.31	1.3
C_FA	60	0.66	0.5	0.034	0.26	0.87	2.1

Wilk test), Pearson's correlation coefficient was used in this example (Fig. 3.3). Other assumptions that should be followed to correctly apply a Pearson correlation include the following (Schober et al. 2018): (**1**) the data are a random or representative sample of the population; (**2**) both variables are continuous, jointly normally distributed, random, and follow a bivariate normal distribution of the population from which they were sampled; (**3**) the relationship between the data is linear; (**4**) there are no significant outliers; and (**5**) each pair of XY values is measured independently from other pairs (Schober et al. 2018). Notably, in many cases, we do not numerically confirm that the data follow the normal distribution but rather find that their distribution is close to normal. On the basis of the **Central Limit Theorem** (CLT), we can assume that the assumption of normality is valid. Following the CLT, we assume that the sample distribution increasingly approximates a normal distribution with increasing sample size, even if the population distribution is not normal. As shown in Fig. 3.3, two pairs of variables are strongly correlated: N with Cox ($r = 0.99$) and Ca with Mg ($r = 0.92$). These correlations are statistically significant, as indicated by three red stars over the correlation values (in this case, the p value is <0.0001). At the same time, there are other variables for which the correlation is comparatively weaker, but the relationship is still statistically significant, such as Cox and Ca ($r = 0.8$). The function description does not inform us of what method was used to calculate the significance of the relationship. However, the user can press the F2 key with the cursor over the function name to see the function source code in a new RStudio window. Note that this process is not the same as requesting information from the Help panel, which can be obtained by executing ***?function_name()*** (function name with a question mark). When evaluating significance with the ***chart.Correlation()*** function, the ***cor.test()*** function, which is a part of the **stats** package, is employed (R Core Team 2023). This function is used to perform a correlation test for each pair of variables in the dataset using the t test (Obilor and Amadi 2018). In addition, the scatterplots indicate linear relationships between Cox and N and between Ca and Mg. There is one outlier, but we ignore it at the moment. However, outliers can be the result of measurement error or a typo, and inspecting them before any more complex analyses and modeling is good practice. Most of the variables in our example are normally distributed and can be evaluated on the basis of histograms. The normality can also be tested computationally using the Shapiro–Wilk (SW) test (for example, the ***shapiro.test()*** function) or visually via quantile-quantile (QQ) plots (for example, the ***ggpubr::ggqqplot()*** function) (Fig. 3.4). The result of the SW test is the W statistic, which varies between zero and one. Small values of W indicate a lack of normality, and values close to one indicate normality (Razali and Wah 2011).

With respect to the statistical significance level, controversies exist with respect to p values, which were formulated by the American Statistical Association (ASA) (Ronald et al. 2016). Among several recommendations, the authors propose the following:

1. Scientific conclusions and business or policy decisions should not be based only on whether a p value reaches a specific threshold.
2. By itself, a p value does not provide a good measure of evidence regarding a model or hypothesis.

```
pacman::p_load("PerformanceAnalytics")
#We will start by testing numeric variable
correlations using the PerformanceAnalytics
package.
#Loading the package.
library(PerformanceAnalytics)
#We select only eight variables from the
dataset for a better visualization. The chart.
Correlation() function performs the Pearson's
correlation calculation (default option). Other
methods that can be selected include Kendall's
and Spearman's correlation coefficients.
#For this task, we assume the normal
distribution of the variables, but the
normality can also be tested numerically and
graphically.
shapiro.test(gorce_soils$pH_H2O) #select one
variable
#The data is normally distributed if the W
statistic is near zero and the p-value is over
0.05. We can also evaluate it by inspecting the
QQ plot.
library(ggplot2)
ggplot(gorce_soils, mapping = aes(sample =
pH_H2O))+
  stat_qq()+
  stat_qq_line()+
  ggtitle(expression("QQ plot for soil
pH"[H2O]))
#ggsave("ggplot.png")
```

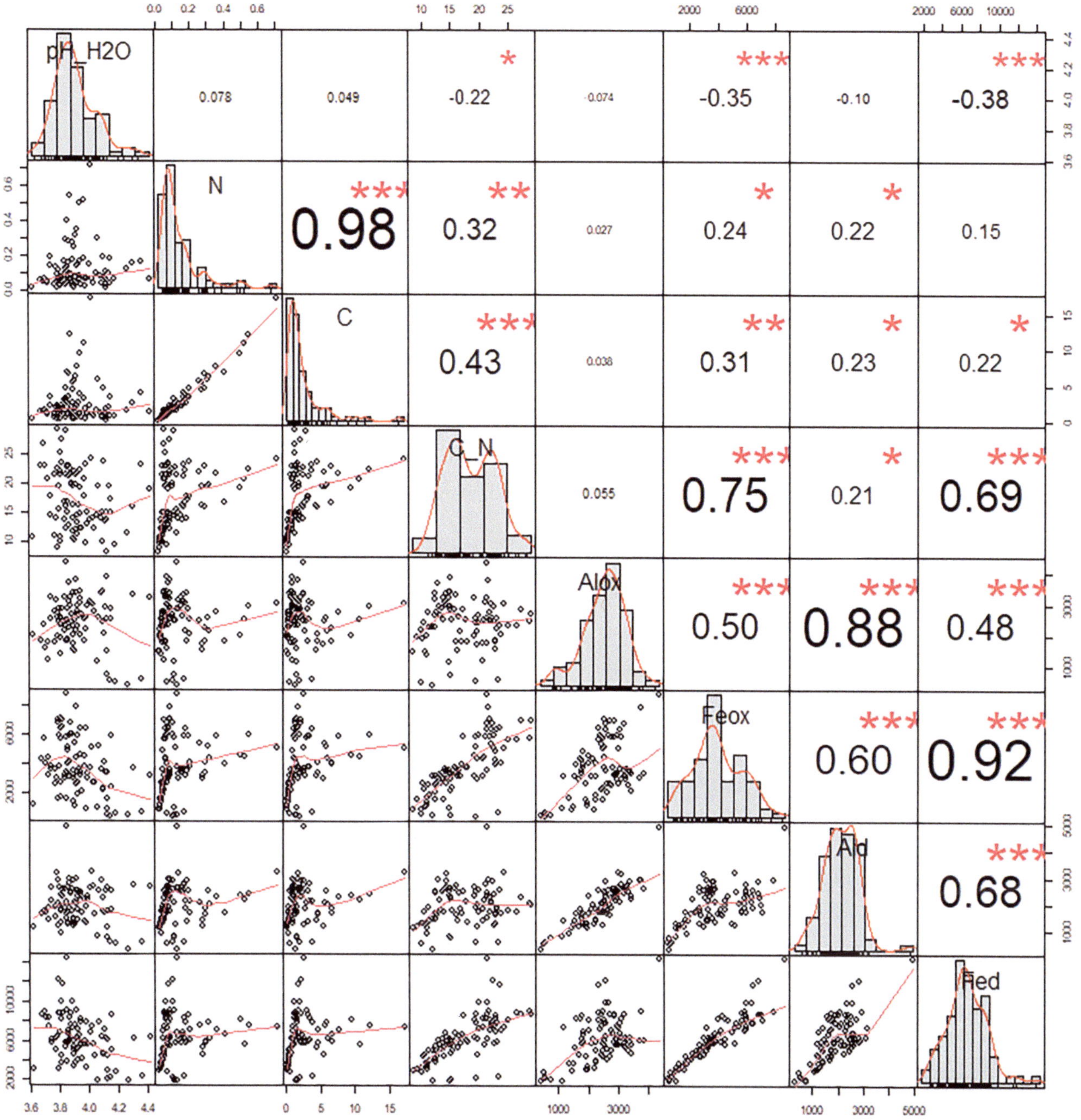

Fig. 3.3 A correlation chart consisting of scatter plots, histograms, the calculated correlation coefficients, and their significance for each pair of variables. The figure was prepared using the **PerformanceAnalytics** package (Peterson and Carl 2020)

```
#To save the newly created file as PNG, we first
open the PNG device and define the output
directory path and the file name with the
extension.
#png("corr_chart1.png", width = 800, height =
800)
#We select 8 out of 30 variables.
library(dplyr)
gorce_soils[,c("pH_H2O", "Cox", "N", "Ca", "K",
"Mg", "Na", "Ald")] |>
 select_if(where(is.numeric)) |>
 chart.Correlation()
#Closing the graphic device.
#dev.off()
```

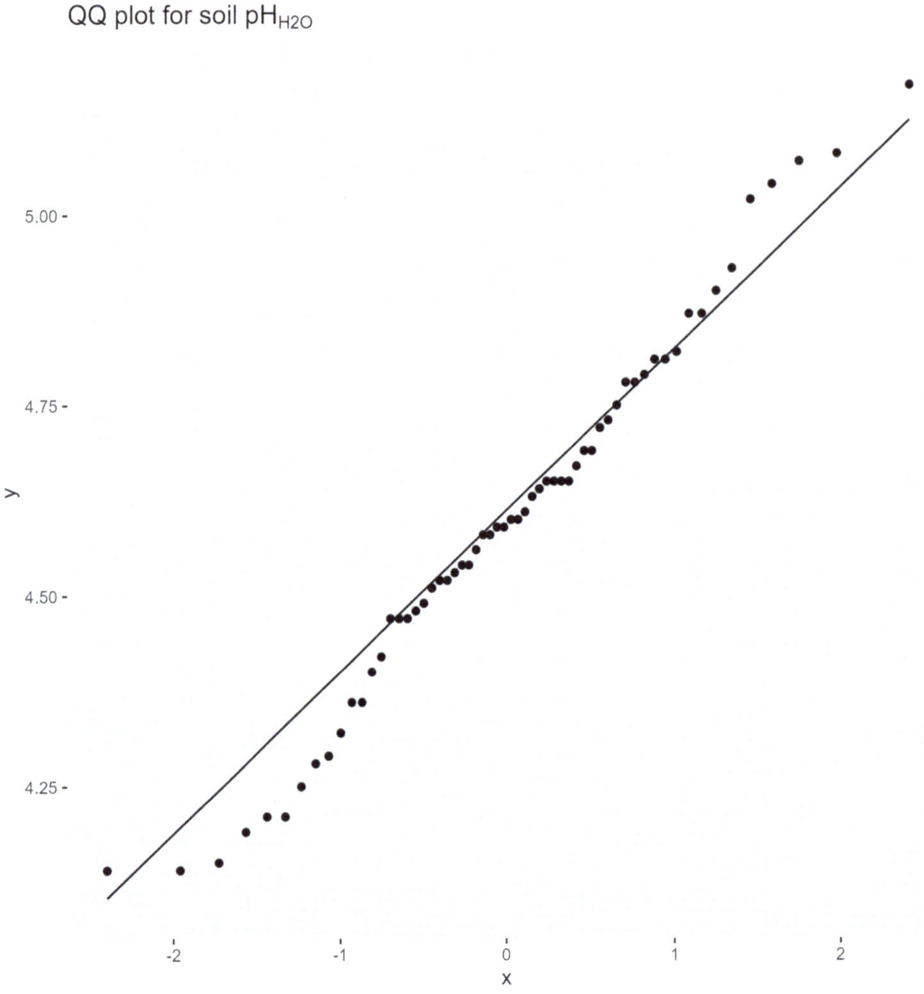

Fig. 3.4 The quantile–quantile (QQ) plot for data normality inspection

Another popular R package that aids in visualizing correlation matrices is the **corrplot** package (Wei and Simko 2021). Figure 3.5 displays the results of Pearson's correlation coefficients for selected soil variables from the Gorce dataset. The colors indicate the correlation direction (−1 indicates a strong negative relationship, and +1 indicates a strong positive relationship), and the size of the colored squares represents the correlation magnitude. The numbers inside the boxes indicate the r values. Note that the correlation matrix is symmetrical along a diagonal; thus, it is sufficient if only the lower or upper part of the matrix is shown.

```
#Testing correlations with the corrplot
package.
pacman::p_load("corrplot")
library(corrplot)
#In this step, a correlation matrix is
calculated.
#We plot it using all continuous variables
without modifying the figure.
corrplot(cor(gorce_soils[,4:30]))
#How can we improve it?
cor1 <- cor(gorce_soils[,c("pH_H2O", "Cox",
"N", "Ca", "K", "Mg", "Na", "Ald")],
 use = "na.or.complete") #with the use
 argument, we select the way the function deals
 with NA values. use = "na.or.complete" means
 that correlations will be computed from
 complete observations, NA observations are
 removed
#To see the difference, run the following
example.
test1 <- data.frame(a = c(1:10),
                    b = c(1:5, NA, 7:10),
                    c = c(21:30),
                    d = c(0.1:1.1))
cor(test1, use = "na.or.complete")
cor(test1)
#In this step, a significance test provides
p-values and confidence intervals for each pair
of input features.
```

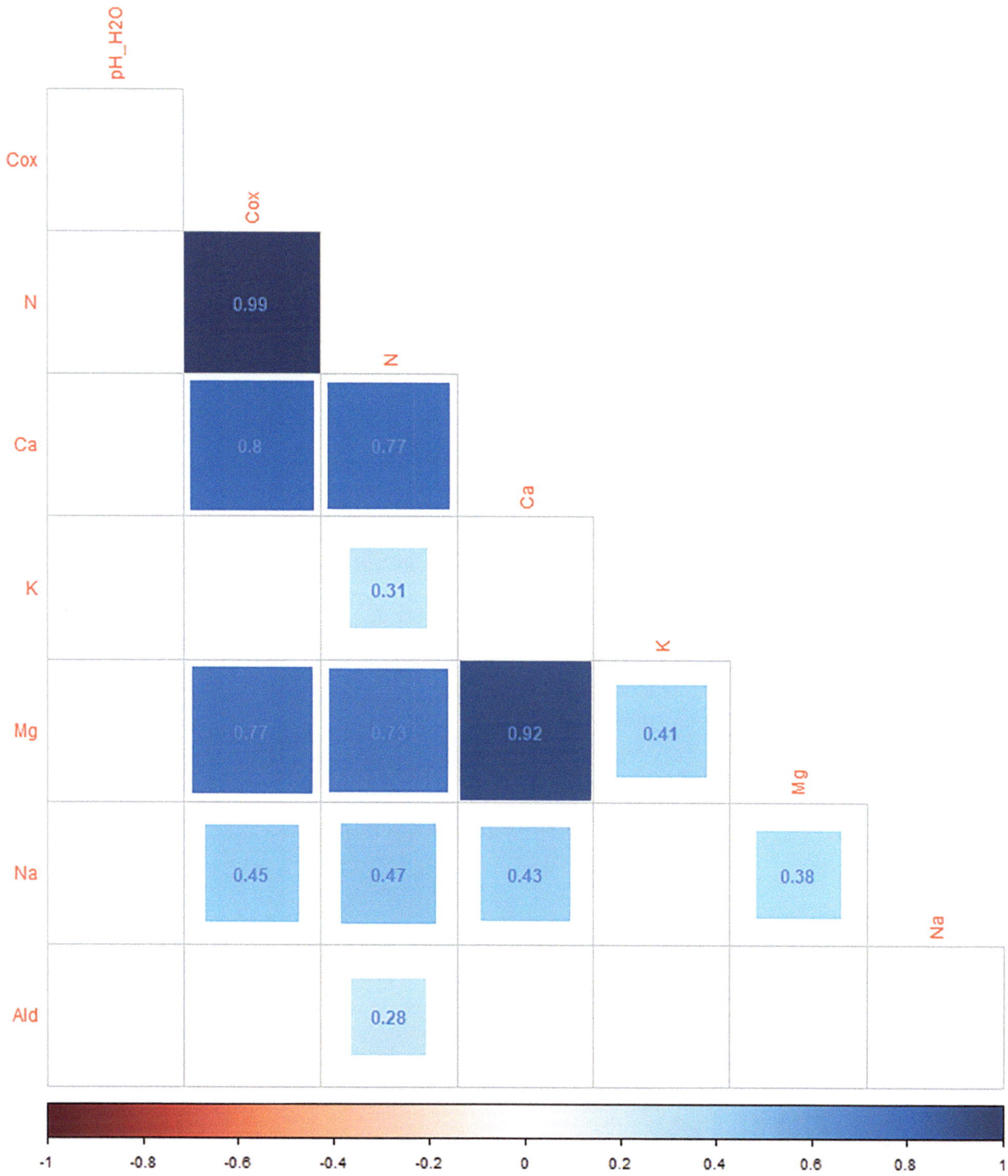

Fig. 3.5 A correlation chart was constructed using the **corrplot** library (Wei and Simko 2021). The cells with nonsignificant correlations were left blank

```
cor_test1 <- cor.mtest(gorce_soils[,c("pH_H2O",
"Cox", "N", "Ca", "K", "Mg", "Na", "Ald")],
conf.level = .05)
#opening the graphic device
#png("corrplot1.png", width = 800, height =
800)
corrplot(cor1, p.mat = cor_test1$p,
            method = "square", type = "lower",
            sig.level = c(0.001, 0.01, 0.05),
            pch.cex = 0.8,
            insig = "blank", addCoef.col =
            'dodgerblue',
            diag=FALSE, number.cex = 1,
      pch.col = "tomato")
#closing the graphic device
#dev.off()
```

As soil analyses are usually complex and the correct interpretation and identification of soil units require information on physical and chemical soil properties, a tabular format with a dense matrix of rows and columns is not always the optimal way to display the final data. A ternary diagram is one tool for elegant data visualization and the most popular solution for the rapid evaluation of soil and sediment particle size distributions. It is commonly used in soil science, sedimentology, geology, and geomorphology. With a ternary diagram, we can obtain a clear picture of the proportions of the three main soil fractions (sand, silt, and clay). However, these synthetic soil data lack information about the depth of the main horizons, which can be added as an additional parameter, with values shown on a continuous color scale (Fig. 3.6). Another option is to use a functionality in the **aqp** package and build a SoilProfileCollection, which is an R object that can be plotted (Fig. 3.7). In this way, the user can evaluate individual soil profiles and their properties along the vertical scale, i.e., for different horizons.

```
pacman::p_load("aqp")
library(aqp) #package for quantitative pedology
#A new object is created called
SoilProfileCollection, which needs top and
bottom variables that indicate the upper and
lower limits of subsequent soil horizons in a
soil profile.
#'id' is the name of the column containing the
profile ID
# 'top' is the name of the column containing
horizon upper boundaries
# 'bottom' is the name of the column containing
horizon lower boundaries
depths(gorce_soils) <- id ~ top+bottom
class(gorce_soils)#check the object class
#Saving several plots as a PNG file.
#png("soil_plots.png", width = 800, height =
600)
```

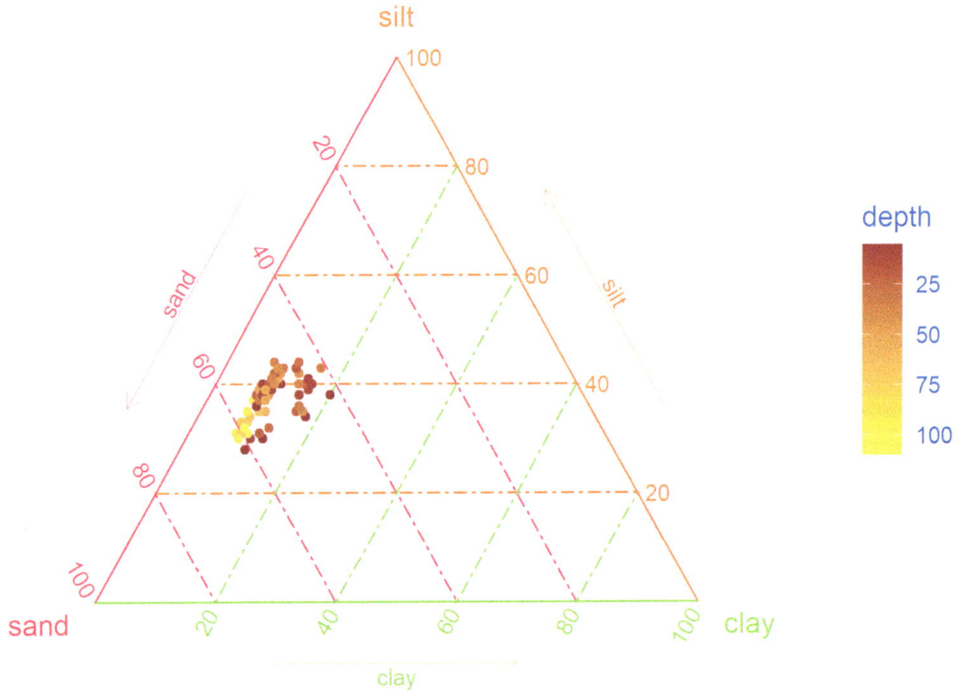

Fig. 3.6 Ternary diagram displaying synthetic information on the main soil fractions of samples collected in Gorce National Park, Poland. The data are from Pawlik and Samonil (2018). In each case, the sand fraction ranges from 40–60%, the silt fraction ranges from 20–40%, and the clay fraction is 0–20%

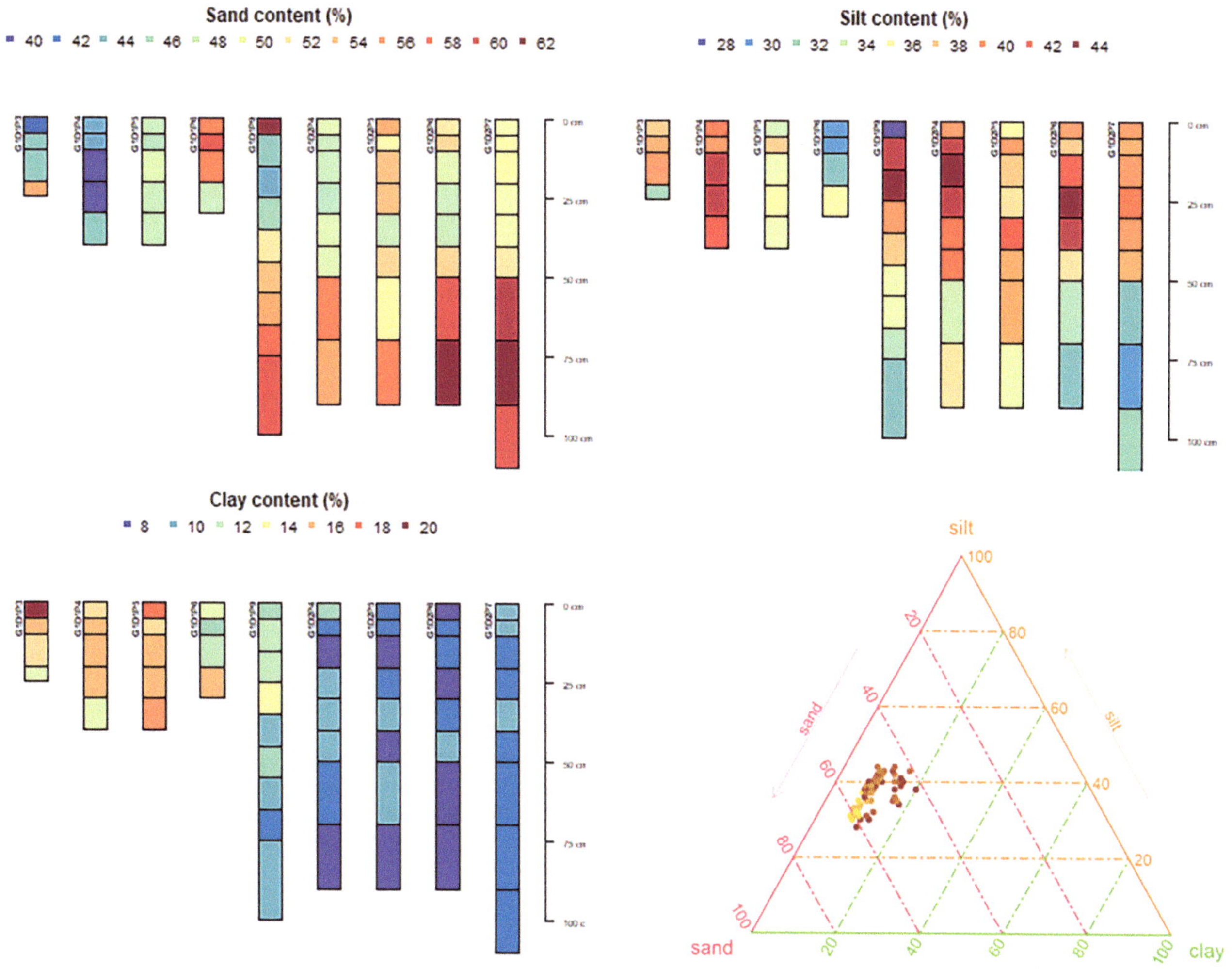

Fig. 3.7 Visualization of the main soil fractions (particle-size distributions) along individual profiles analyzed in Gorce National Park, Poland, compared with a ternary diagram created with the same dataset. Data source: (Pawlik and Samonil 2018)

```
par(mfrow=c(2,2))#setting parameters for the
number of rows and columns for a combination of
three plots in one figure
par(mar=c(0,0,3,1))#setting parameters for the
number of lines of margin (bottom, left, top,
right)
#Three plots combined into one figure.
plot(gorce_soils, color="sand", col.label =
"Sand content (%)", width = .2, id.style =
"side")
plot(gorce_soils, color="silt", col.label =
"Silt content (%)", width = .2, id.style =
"side")
plot(gorce_soils, color="clay", col.label =
"Clay content (%)", width = .2, id.style =
"side")
#dev.off()
library(ggtern)
ggtern(data = as.data.frame(gorce_soils),
aes(x=sand, y=silt, z=clay, color=bottom))+
  geom_point()+
  scale_color_gradient(low="brown", high =
  "yellow")+
  theme_tropical()+
  guides(color = guide_colorbar(reverse =
  TRUE))+
  labs(color = "depth")
#ggtern::ggsave("ternary_chart1_gorce_soils.
png")

#In this part, we will combine plots displaying
soil horizonation and the ternary plot. To do
that, we need to import them as graphical
objects. The magick package will be used to
merge these plots into one figure.
pacman::p_load("magick")
library(magick) #image processing library
#Read the image from the folder in which it was
saved. Modify the file directory in the image_
read() function to import the image.
```

```
soil_hor <- image_read("soil_plots.png")
#Note: the image can also be read from the
GitHub repository. Activate the code below by
removing the # sign,= if you want to use it
this way.
#soil_hor <- image_read("https://raw.github.
com/Lukpawlik/SN_figures/main/soil_plots.png")
#Read the image from the folder in which it was
saved. Modify the file directory in the image_
read() function to import the image.
ternary_soils <- image_read("ternary_chart1_
gorce_soils.png") |>
 image_scale("550x") #image scaling
#Note: the image can also be read from the
GitHub repository. Activate the code below by
removing the # sign,= if you want to use it
this way.
#ternary_soils <- image_read("https://raw.
github.com/Lukpawlik/SN_figures/main/ternary_
chart1_gorce_soils.png") |> image_scale("550x")
#image scaling
#Combine images.
image1 <- image_composite(soil_hor, ternary_
soils, offset = "+390+290")
image1 #plot the image
#Save the image.
image_write(image1, "combined_soil_images.png")
```

References

Arel-Bundock V (2022) Modelsummary: data and model summaries in R. J Stat Softw 103:1–23. https://doi.org/10.18637/jss.v103.i01

Beaudette D, Roudier P, Brown A (2022) aqp: algorithms for quantitative Pedology. R package version 1.42

Corral-Pazos-de-Provens E, Rapp-Arrarás Í, Domingo-Santos JM (2022) Estimating textural fractions of the USDA using those of the international system: a quantile approach. Geoderma 416:115783. https://doi.org/10.1016/j.geoderma.2022.115783

Dubeux JCB, de Mário ALJ, Simili Flávia F et al (2024) Deep soil organic carbon: a review. CABI Rev 19:1. https://doi.org/10.1079/cabireviews.2024.0024

Hamilton NE, Ferry M (2018) ggtern: ternary diagrams using ggplot2. J Stat Softw 87:1–17. https://doi.org/10.18637/jss.v087.c03

Howarth RJ (1996) Sources for a history of the ternary diagram. Br J Hist Sci 29:337–356. https://doi.org/10.1017/S000708740003449X

Huntington-Klein N (2023) vtable: variable table for variable documentation

IUSS Working Group WRB (2022) World Reference Base for soil resources. In: International soil classification system for naming soils and creating legends for soil maps, 4th edn. International Union of Soil Sciences (IUSS), Vienna

Kellogg CE (1941) The Soils That Support Us, The Macmillan Company, New York City, 370 pp

Obilor EI, Amadi EC (2018) Test for significance of Pearson's correlation coefficient (r). Int J Innov Math Stat Energy Policies 6:11–23

Pawlik Ł, Kasprzak M (2018) Regolith properties under trees and the biomechanical effects caused by tree root systems as recognized by electrical resistivity tomography (ERT). Geomorphology 300:1–12. https://doi.org/10.1016/j.geomorph.2017.10.002

Pawlik Ł, Samonil P (2018) Biomechanical and biochemical effects recorded in the tree root zone—soil memory, historical contingency and soil evolution under trees. Plant Soil 426:109–134. https://doi.org/10.1007/s11104-018-3622-9

Pawlik Ł, Gruba P, Gałązka A et al (2023) Weathering and soil production under trees growing on sandstones—the role of tree roots in soil formation. Sci Total Environ 902:166002. https://doi.org/10.1016/j.scitotenv.2023.166002

Pawlik Ł, Gałązka A, Gruba P et al (2024) High-resolution soil sampling reveals the pattern of biological weathering and soil formation under trees. Sci Total Environ 941:173725

Peterson BG, Carl P (2020) Performance Analytics: econometric tools for performance and risk analysis

Pohlmann A (2020) Measures and numbers for colors. The color system of Wilhelm Ostwald. ChemTexts 6:9. https://doi.org/10.1007/s40828-020-0104-5

Post WM, Pastor J, Zinke PJ, Stangenberger AG (1985) Global patterns of soil nitrogen storage. Nature 317:613–616. https://doi.org/10.1038/317613a0

R Core Team (2023) R: a language and environment for statistical computing

Razali NM, Wah YB (2011) Power comparisons of Shapiro-Wilk, Kolmogorov-Smirnov, Lilliefors and Anderson-Darling tests. J Stat Model Anal 2:21–33

Ronald L., Wasserstein Nicole A., Lazar (2016) The ASA Statement on p -Values: Context Process and Purpose. The American Statistician 70(2):129–133 https://doi.org/10.1080/00031305.2016.1154108

Schaetzl RJ, Thompson ML (2015) Soils—genesis and geomorphology, 2nd edn. Cambridge University Press, Cambridge

Schober P, Boer C, Schwarte LA (2018) Correlation coefficients: appropriate use and interpretation. Anesth Analg 126:1763–1768. https://doi.org/10.1213/ANE.0000000000002864

Smith MR (2017) Ternary: an R package for creating ternary plots

Wei T, Simko V (2021) R package "corrplot": visualization of a correlation matrix

Wickham H (2016) ggplot2: elegant graphics for data analysis. Springer, New York

Yongue-Fouateu R, Ndimukong F, Njoya A et al (2016) The Ndop plain clayey materials (Bamenda area—NW Cameroon): mineralogical, geochemical, physical characteristics and properties of their fired products. J Asian Ceramic Soc 4:299–308. https://doi.org/10.1016/j.jascer.2016.05.008

4 Geochemistry of Fossil Deposits and Climate Reconstruction Based on Proxy Data

Serra da Canastra National Park, Brazil, 2024, photo: Ł. Pawlik

> There we found the same continuity and perfect preservation of the layers clearly representing annual sediment deposits.
>
> Gerard de Geer (1910, 1912), translated by Christian Dullo and William W. Hay (de Geer 2002)

In Quaternary science and other branches of Earth sciences, **proxy data** provide indirect information on the climate or environmental conditions of the past (Walker 2005; Mauri et al. 2015). Because scientists do not have the chance to directly measure or observe the properties of past climates or environmental changes (for example, wildfires or soil erosion) in detail, other sources of data need to be explored and combined in support of any reconstruction and final conclusions. Proxy indicators are interpreted in light of our knowledge of the present conditions and their impact on the biosphere, pedosphere, and landscape dynamics. These indicators are also compared with other proxy data to identify notable similarities or striking differences. A valuable source of information is determining places where various types of materials could have been deposited and kept intact. This means that the deposits formed a set of layers that were not disturbed during or after deposition, did not contain gaps (hiatuses), and were not physically or chemically contaminated during or after deposition. The most favorable conditions for this type of continuous deposition are in ocean and sea basins, at lake bottoms, and in ice bodies (glaciers, ice

Supplementary Information The online version contains supplementary material available at https://doi.org/10.1007/978-3-031-89673-6_4.

© The Author(s), under exclusive license to Springer Nature Switzerland AG 2025

Ł. Pawlik, *R Applications in Earth Sciences*, Springer Textbooks in Earth Sciences, Geography and Environment,
https://doi.org/10.1007/978-3-031-89673-6_4

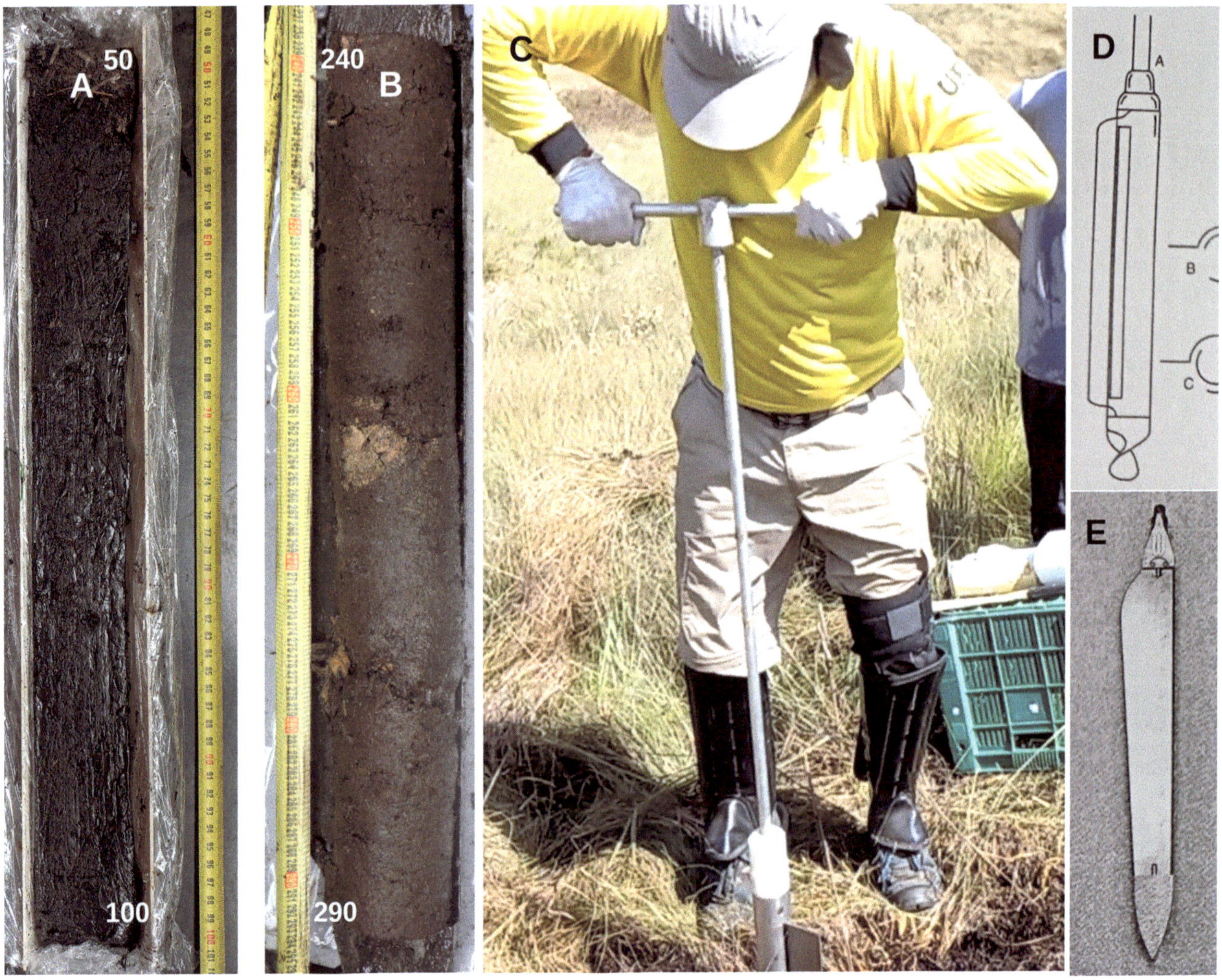

Fig. 4.1 Examples of cores with decomposed peat sediments from the Serra da Canastra National Park, Brazil (**a**), and the Wyspowy Beskidy Mountains, Poland (**b**). The cores were collected using a handheld borer (**c**). Two examples of borers are shown: the Hiller peat borer (**d**) and the Instorf peat borer (**e**) (after Grażyna Miotk-Szpiganowicz). The depth is given in cm

fields, and ice caps). Other natural data archives include tree rings and karst features (for example, stalactites and stalagmites). These features are potential sources of highly valuable information on past climate and environmental conditions. This information is obviously incomplete and indirect, but at the same time, it is unique and often the only data we have available (Fig. 4.1).

4.1 Description of the Data and Project Outline

We focus on a dataset including information derived from thin (2 cm) layers of a 4-m-deep core. This core was collected from a landslide fen that formed on the northern part of Mt. Śnieżnica (1006 m asl) in southern Poland. This landslide is estimated to be at least 14,000 years old (Pawlik et al. 2022; section 7). A large amount of soil slid from a mountain slope. This material formed a landslide body with a highly irregular surface topography and numerous pits and depressions that are seasonally dry or permanently filled with water (Fig. 4.2). These places are perfect for the accumulation of mineral and organic material, which has been stored there for thousands of years. This material was excavated with the use of a hand auger and analyzed in a laboratory to extract information about the contents of numerous elements and chemical compounds. The final dataset contains several variables, and their measured values are expressed in %, amount per 1 cm^3, mg g^{-1}, or ppm (µg g^{-1}). These measurements were interpreted in a certain environmental context and provide a source of information regarding regional climate and land cover changes during the past 5000 years.

Fig. 4.2 Study area where sediments were collected from a pit within a landslide fen (yellow star). Note the irregular, hilly topography north of the main ridge of the mountain and the clear distinction between the landslide area and undisturbed area, which is characterized by a flat and smooth surface. The landslide body is very complex and forms along a clearly visible scar with a NW–SE orientation

4.2 R in Practice: Geochemical Data Analysis and Visualization

During every project, several tools and methods will be needed and applied, starting from file import and read functions and perhaps finishing with final plot and its interpretation. In the final stage, the data are explored, transformed, modeled, and summarized. In many instances, the functionalities of at least three or four packages are used to achieve the desired goal. In this project, we first read the data, calculate descriptive statistics, and check the data structure.

```
#Packages used in this section.
if(!require("pacman")) install.
packages("pacman")
#The pacman package is used to check if we need
to install the required packages (Rinker and
Kurkiewicz 2017).
pacman::p_load("dplyr", "ggplot2", "reshape2",
"ggpubr",
               "ggsignif", "ggstatsplot",
               "rstatix", "hablar", "readr")
library(dplyr) #data manipulation
library(ggplot2) #visualization and plotting
library(reshape2) #data reshaping
library(ggpubr) #publication ready plots
library(ggsignif) #data comparison
library(ggstatsplot) #plots with details from
statistical tests
library(rstatix) #statistical tests
library(hablar) #converting columns to new data
types
library(readr) #read rectangular data
#Table with the measured geochemical variables
was saved in a CSV file
#The table is a part of the rappliES package.
library(rappliES)
data(geochemistry)
```

```
#Explore the data.
summary(geochemistry) #descriptive statistics
str(geochemistry) #data structure
names(geochemistry) #variable names
```

The main task is to analyze the distribution of each variable along the vertical profile, from the ground surface to the deepest part of the core. For this task, line graphs were used. Because the content of each variable was determined along the same profile, one solution was to use the ***facet_wrap()*** function from the **ggplot2** library. With this function, a fraction of the dataset is displayed in each subplot, and subplots are arranged in columns and rows (Fig. 4.3). Before the function was used, we changed the data frame from wide (all variables as columns) to long (variable names in one column) using the ***melt()*** function from the **reshape2** package (Wickham 2007). With the ***ggplot2::geom_path()*** function, instead of the ***ggplot2::geom_line()***, the observations were connected in the order in which they appeared in the data. In this way, and by using the ***ggplot2::scale_y_reverse()*** function, the desired effect of the correct position of variable values along the Y axis was obtained. By examining the plot, striking changes can be observed in the accumulation of calcium (Ca), carbon (C), and, for instance, magnesium (Mg). A similar pattern of changes along the profile is observed for sodium (Na), potassium (K), and Mg, with decreasing values from the bottom of the profile to a depth of two meters. These elements are highly correlated and are subsequently examined via a principal component analysis (PCA) biplot. Sulfur (S) and C display different fluctuating trends in general, but below a depth of two meters, their contents are similar. Moreover, the plot of the phosphorus (P) content has a very distinct shape.

In the next code chunk, we apply a pipe operator to connect two different processes: (1) data manipulation–filtering and (2) plotting. This operator allows us to shorten the code without creating new object. Please note that the pipe operator **%>%** is changed to **+** when executing the **ggplot2** part of the code.

```
#You will use the melt() function to modify the
table. The function is from the reshape2 R
package.
geochemistry_melt <- melt(geochemistry, id.vars
= 'Depth')
head(geochemistry_melt)#What do you observe
after calling the object?
#Note that all variables were reduced to one
column, and all values are in the "value"
column. This operation allows the facet_wrap()
```

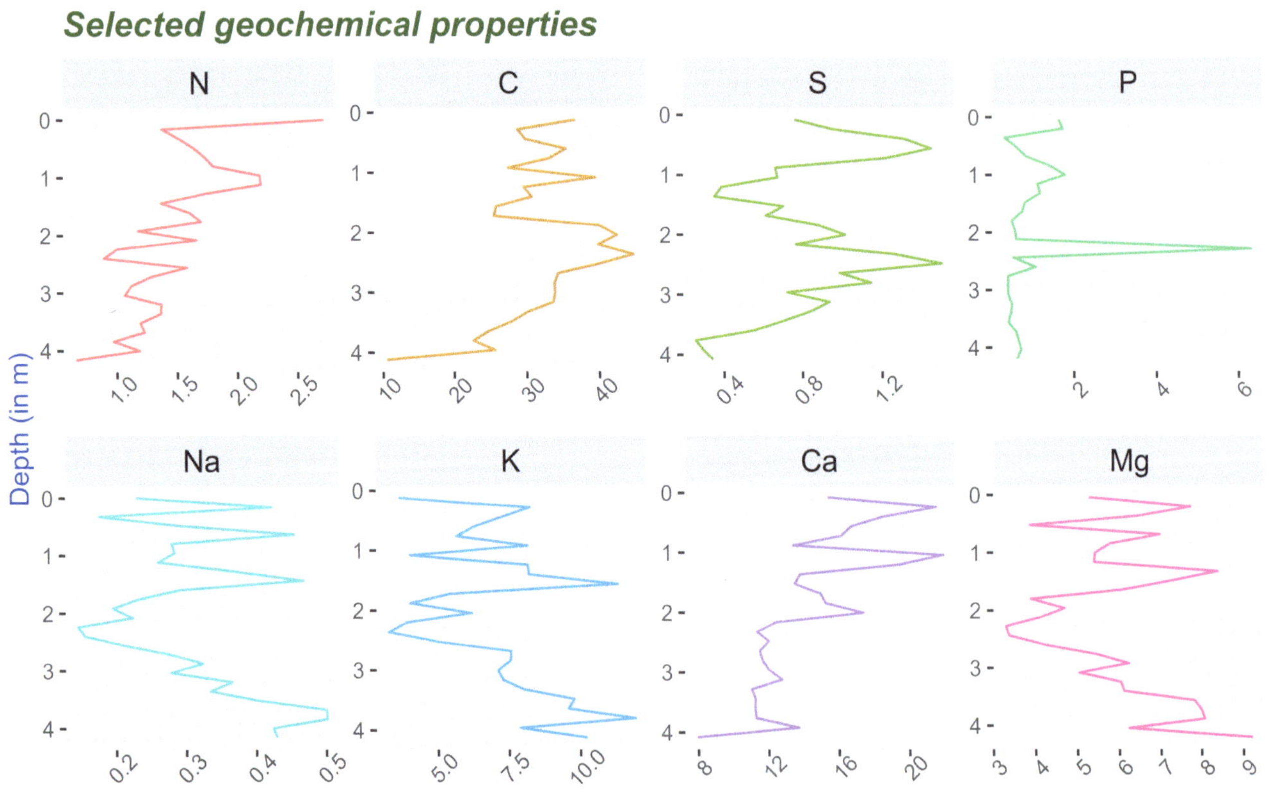

Fig. 4.3 Reversed vertical line plots with information on changes in the basic chemical properties of the sediments across the depths of the samples. Data source: Pawlik et al. (2022)

```
function from the ggplot2 package to be
applied.
geochemistry_melt %>% filter(variable == c("N",
"C", "S", "P", "Na", "K", "Ca", "Mg")) %>%
 ggplot()+
 geom_path(aes(x=value, y=Depth,
color=variable))+
 scale_y_reverse()+
 facet_wrap(~variable, nrow = 2, scales =
'free')+
 ylab("Depth (in m)")+
 ggtitle("Selected geochemical properties")+
 theme
(plot.title = element_text(color="darkgreen",
size=14, face="bold.italic"),
 axis.text.x = element_text(angle=45),
 axis.title.y = element_text(color = "blue"),
        legend.position = 'none',
        axis.title.x = element_blank(),
        strip.text.x = element_text(size = 12))
#The plot can be saved as a PNG file.
#ggsave("geochemistry_line_plots.png", dpi=600)
```

In the next steps, the dataset is evaluated by employing PCA, an unconstrained ordination method used for rapid evaluation of the data structure, correlations between variables, and the proportion of data variance explained by each principal component. PCA was introduced by Pearson (1901) and is used to reduce the dimensionality of multivariate data when the core components are extracted from multiple variables after orthogonal transformation (Daróczi 2015; Borcard et al. 2018; Lesmeister and Chinnamgari 2019; Boehmke and Greenwell 2020). This technique is used when many variables are closely correlated, which means that they are associated with a similar amount of information (changes in variable A mirror changes in variable B) that can be retained even if some set of variables is reduced to new uncorrelated components. The goal is to create a dataset with a small number of variables (principal components) that capture a large part of the information from the original dataset (Lesmeister and Chinnamgari 2019; Boehmke and Greenwell 2020). A component is a normalized linear combination of variables. The variables can originate from heterogeneous data sources and can be expressed in different units. However, the data should be numeric, without NA values, and standardized (e.g., centered and scaled, especially when variables have different units) to make them comparable (Boehmke and Greenwell 2020). The principal components are uncorrelated, and the first principal component (PC1) explains the largest amount of variation (Lesmeister and Chinnamgari 2019). How are the components determined? In a two-dimensional space (with just two variables), the line passes through the scatter plot in such a way that it minimizes the total squared distance from each point to its orthogonal projection onto the line (Boehmke and Greenwell 2020). The second principal component (PC2) is orthogonal to PC1. Figure 4.4 shows how a PCA biplot is constructed and how it is used to map data to a low-dimensional space while preserving as much information as possible (Friendly and Sigal 2014). The name "biplot" underscores the fact that both the observations and the variables are displayed in a single plot (Gower et al. 2011; Friendly and Sigal 2014). A standard PCA biplot contains component scores (points) and the loadings of variables (arrows and vectors). On the basis of the length and direction of the variable loadings, we can assess their relationships with each of the components and their contributions to the principal components PC1 and PC2.

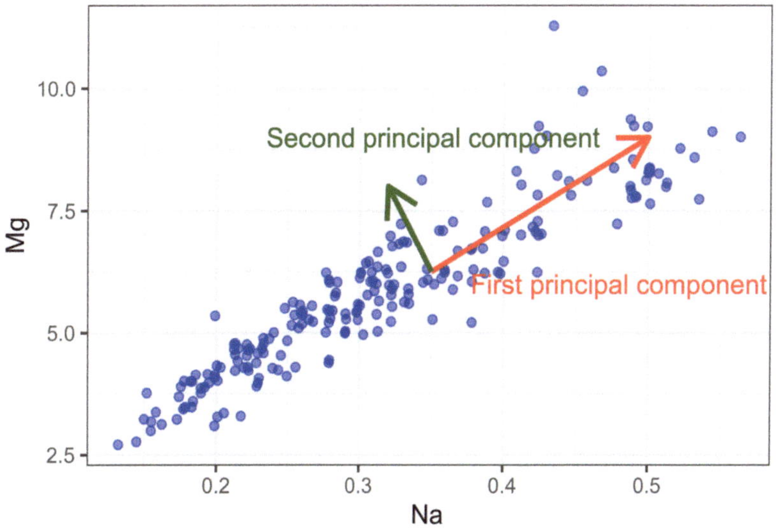

Fig. 4.4 Scatterplot showing the relationship between two variables and how a PCA biplot is constructed with the first and second principal components isolated. Please note that the axes are perpendicular to each other, and in the resulting PCA biplot, they are rotated, as shown in the next figure

```
#Scatter plot with two variables, Na and Mg.
ggplot(geochemistry)+
 geom_point(aes(x=Na, y=Mg), color="blue",
 alpha=0.5)+
 geom_segment(aes(x=.35, y=6.25, xend=.5,
 yend=9), arrow = arrow(length=unit(.5, 'cm')),
 color='red', lwd=1)+
 geom_segment(aes(x=.35, y=6.25, xend=.32,
 yend=8), arrow = arrow(length=unit(.5, 'cm')),
 color='darkgreen', lwd=1)+
 geom_text(label = "First principal component",
 x=0.48, y=6, color="red", size=4)+
 geom_text(label = "Second principal
 component", x=0.35, y=9, color="darkgreen",
 size=4)+
 theme_bw()
#ggsave("scatterplotNaMg.png", dpi=800,
width=12, height=8, units = "cm")
```

In Fig. 4.5 one grouping variable, the upper or lower layer of the analyzed peat profile, was introduced to test the impact of deposition depth on the sediment geochemical composition. We see that some variables can be used for the differentiation of those two layers through component PC2, such as Cu/Zn, TOC/N, $CaCO_3$, and Ca. For the PCA, we used the ***stats::prcomp()*** function; however, there are other popular R packages supporting ordination methods, e.g., the **vegan** package (Oksanen et al. 2018). The first principal component (PC1) explained 46.6% of the variance in the data (Fig. 4.5). The final PCA biplot is visualized with the ***ggbiplot()*** function from the **ggbiplot** package (Vu and Friendly 2024), but there is also a more interactive method of PCA visualization based on the **plotly** package (Sievert 2020) (Fig. 4.6). By examining the plot, we can determine which variables are positively or negatively correlated and which are uncorrelated according to the pattern shown in Fig. 4.7. Variables such as Cu/Zn and Fe/P are related to the characteristics of the lower part of the profile, and Ca and N are related to the

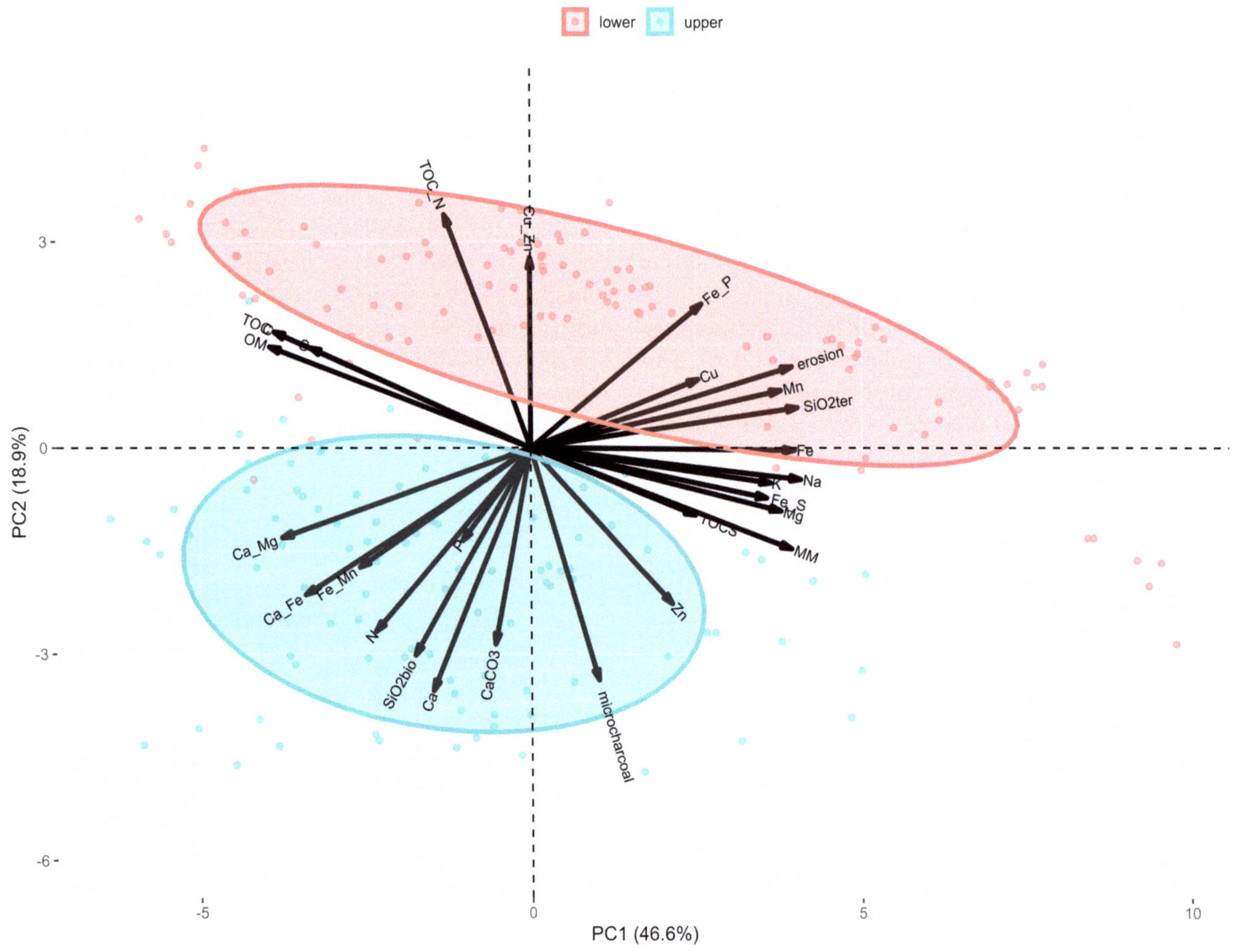

Fig. 4.5 PCA biplot showing geochemical properties measured for samples collected from the core excavated from the landslide fen formed at Mt. Śnieżnica, S Poland. A grouping variable is the depth indicated by two classes: lower—the depth below 2 m, and upper—the depth over 2 m

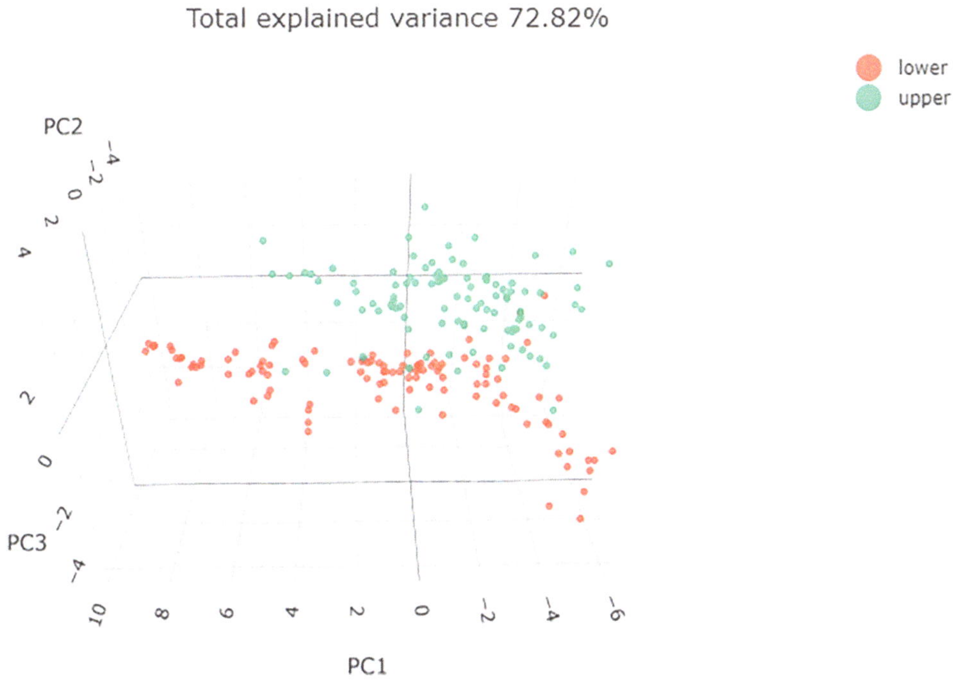

Fig. 4.6 The first three principal components and the level of variance explained by them. The plot style and functionality come from the **plotly** package (Sievert 2020)

characteristics of the upper part of the profile. Moreover, there are positively and negatively correlated variables. For example, the erosion parameter and manganese (Mn) content are positively correlated, and the number of microcharcoals and the TOC/N ratio are negatively correlated. We can also easily identify some noncorrelated variables, such as Zn and Fe/P (Fig. 4.5).

```
pacman::p_load("ggbiplot")
library(ggbiplot)
#Feature engineering - we introduce the depth variable with two classes: "lower" and "upper". They indicate the lower and upper part of the landslide fen profile. For this task, we use the ifelse() statement.
geochemistry1 <- geochemistry %>%
 mutate(layer = ifelse(Depth >=2, "lower", "upper"))
names(geochemistry1)#variable names
#Compute PCA and get the prcomp class object with the following information (please see the function description for more details): sdev (the standard deviations of the principal components), rotation (the matrix of variable loadings), x (the rotated data).
geochemistryPCA <- prcomp(geochemistry1[,c(-1, -31)], scale.=TRUE)
#Plotting.
ggbiplot(geochemistryPCA, obs.scale = 1, var.scale = 1,
        alpha = 0.5, groups = geochemistry1$layer, ellipse = TRUE)+
  theme
(legend.direction = 'horizontal', legend.position = 'top',
        legend.title = element_blank())+
  geom_vline(xintercept = 0, linetype="dashed")+
  geom_hline(yintercept = 0, linetype="dashed")+
  ylim(-6,5)
#Save the file.
#ggsave("pca_geochemistry.png", dpi=600, width = 10, height = 8)

pacman::p_load("plotly")
library(plotly)
#this part was inspired by https://plotly.com/r/
#From the prcomp object, select only the values of the rotated data (the centered and scaled data multiplied by the rotation matrix).
geochemistryPCA1 <- geochemistryPCA[["x"]]
#What will be calculated if we apply the summary() function to the prcomp object?
sum_prcomp <- summary(geochemistryPCA)
sum_prcomp#We get the importance of components.
#Because we want to show three principal components on the interactive plot, we select cumulative proportion up to the PC3 and
```

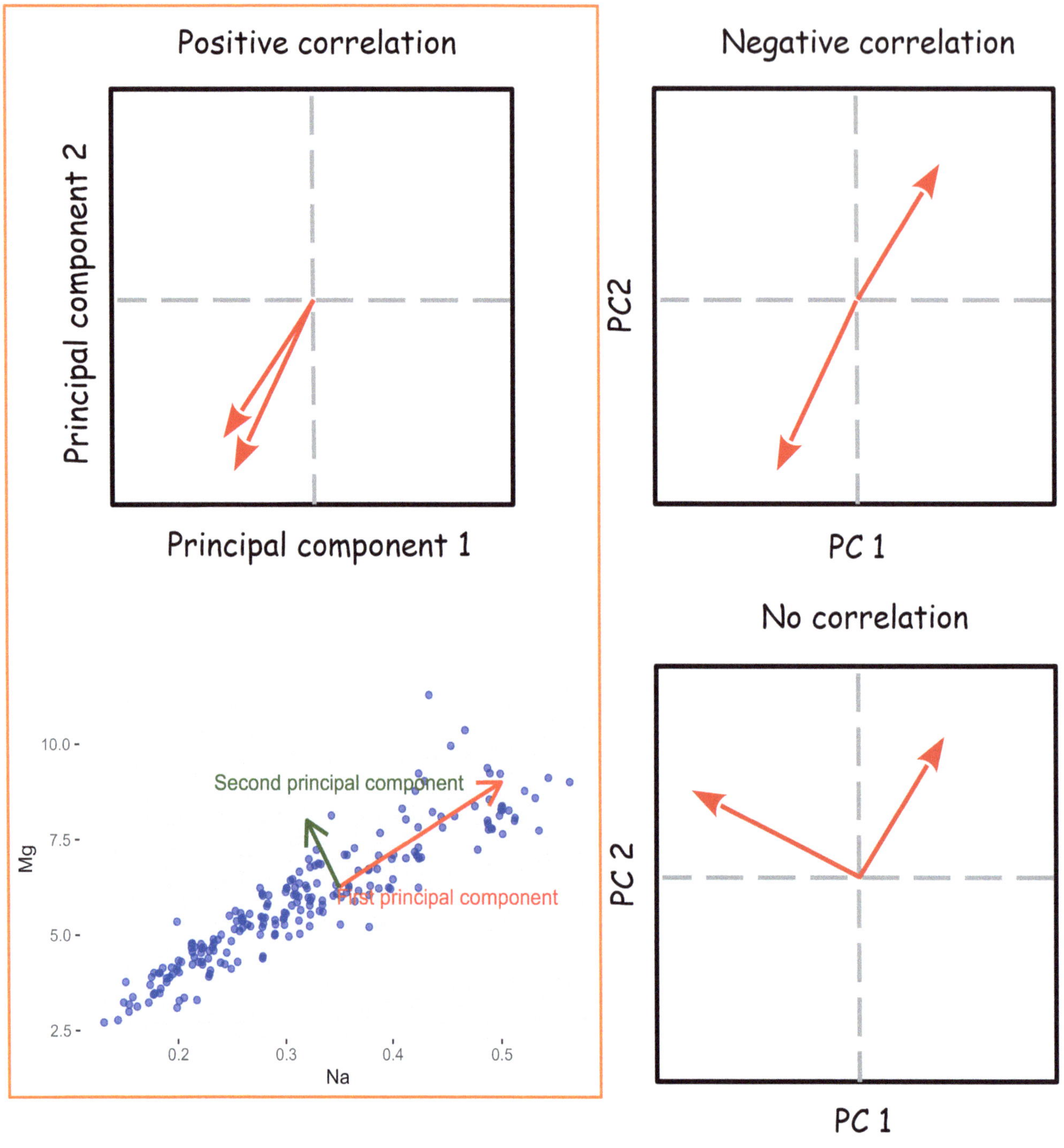

Fig. 4.7 Potential relationships between variables and their interpretation based on the PCA biplot. An example of a positive relationship between Na and Mg and the way principle components are determined when only two variables are analyzed (orange rectangle). Figure inspired by Bertrand et al. (2024) and Boehmke and Greenwell (2020)

```
multiply it by 100. This information will be
used in the plot title (see below).
proportion <- summary(geochemistryPCA)
[['importance']]['Cumulative
Proportion','PC3']*100
#Make sure the data is a data frame object.
geochemistryPCA1 <- data.
frame(geochemistryPCA1)
#Combine the table with information on the
profile depth (the upper and lower layer class).
geochemistryPCA1 = cbind(geochemistryPCA1,
geochemistry1$layer)
#Plotting.
PCAplot <- plot_ly(geochemistryPCA1, x = ~PC1,
y = ~PC2, z = ~PC3, color =
```

```
~geochemistry1$layer, colors =
c('#EF553B','#00CC96') ) %>%
 add_markers(size = 12)
#Add the plot title.
PCAplot <- PCAplot %>%
  layout(
    title = paste0('Total explained variance
',proportion,'%')
  )
PCAplot
```

In the next step, we evaluate the differences between the upper and lower parts of the peat profile below the landslide. As shown by a set of analyses reported by Pawlik et al. (2022), the profile sediments indicate different environmental and depositional conditions between the lower 2-m layer and the upper 2-m layer. Within the history of this reservoir, there was a shift from peat to lacustrine (lake) sedimentation. This interpretation was based on the evaluation of the physical and chemical properties of the material. However, within the next step, we attempt to determine how large the differences between the upper and lower profile segments are and whether there are statistically significant differences between those two layers in terms of the studied variables. We select only four variables to exemplify how differences between categories can be calculated and visualized. The first two variables, the Mg and organic matter (OM) contents, according to the PCA plot (Fig. 4.5), do not impact the differentiation between the upper and lower layers. Their loadings (vectors) are between two ellipses and are closer to the first principal component. The second pair, the TOC/N ratio and the $CaCO_3$ content, has greater contributions to PC2, and in this case, they might add to the difference between the two layers. We tested the significance of differences between layers for each variable. Welch's t test, which is a parametric test for two groups, was used. Welch's test is a t test corrected for unequal variances between samples and is considered a robust alternative to the independent samples t test (Derrick et al. 2016). As expected, there was no significant difference between the groups for Mg ($p = 0.55$) and OM ($p = 0.7$), but there was a significant difference between the groups for TOC/N ($p < 0.0001$) and $CaCO_3$ ($p < 0.0001$) (Fig. 4.8). The effect size was measured on the basis of ***Hedges' g*** metric (Heywood et al. 1998), which is an adjusted (for small group sizes) version of ***Cohen's d*** metric (Clarke 2013). The effect size tells us how large the difference is

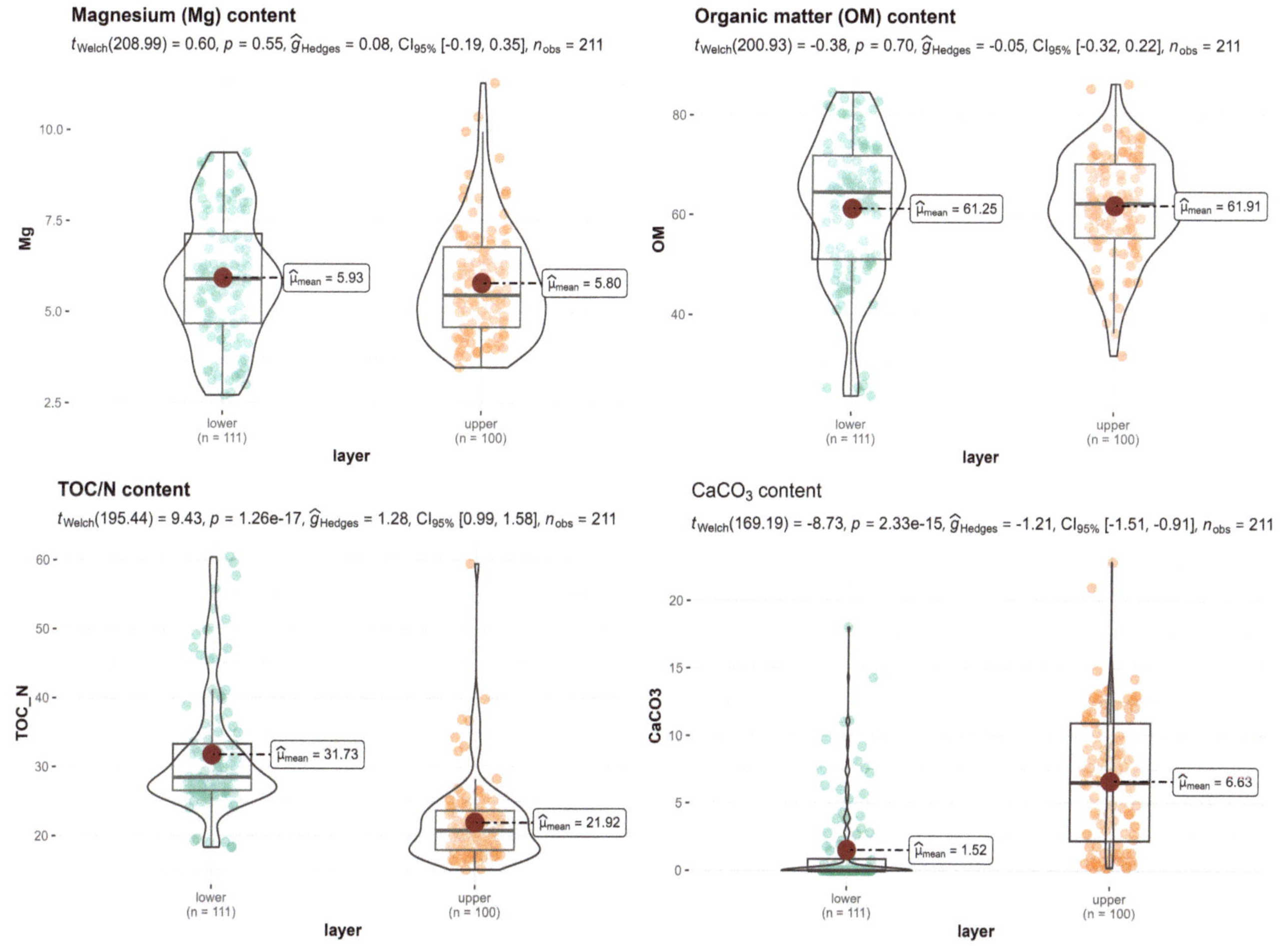

Fig. 4.8 Differences between the upper and lower layers of the analyzed profile for the Mg, organic matter, TOC/N, and $CaCO_3$. A statistically significant difference was detected for TOC/N and $CaCO_3$. The figure was created with the **ggstatsplot** package (Patil 2021)

between groups because even if statistically significant, it can be quite small, and we need a proper measure to verify it. The accepted effect size thresholds for Hodges' g metric are **g = 0.2** (small effect; a small difference between groups), **g = 0.5** (medium effect; a moderate difference), and **g = 0.8** (large effect; a meaningful difference between groups) (Heywood et al. 1998). For the examples presented in Fig. 4.8, we observe a large effect size for groups that differ significantly (proved by Welch's test) but a small effect size for groups that are similar.

```
#We combine four plots using the ggarrange() function from the ggpubr package. To complete the task, four objects are created, each under a unique name. We use a pipe to link the dataset with the plotting function.
#Magnesium (Mg)
mg <- geochemistry1 %>%
 ggbetweenstats(x = layer, y = Mg,
 type = "parametric",
 plot.type = "box",
 pairwise.comparisons = TRUE,
 pairwise.display = "significant",
 centrality.plotting = TRUE,
 bf.message = FALSE,
 title = "Magnesium (Mg) content")
#TOC/N
toc_n <- geochemistry1 %>%
 ggbetweenstats(x = layer, y = TOC_N,
 type = "parametric",
 plot.type = "box",
 pairwise.comparisons = TRUE,
 pairwise.display = "significant",
 centrality.plotting = TRUE,
 bf.message = FALSE,
 title = "TOC/N content")
#CaCO3
caco3 <- geochemistry1 %>%
 ggbetweenstats(x = layer, y = CaCO3,
 type = "parametric",
         plot.type = "box",
         pairwise.comparisons = TRUE,
         pairwise.display = "significant",
         centrality.plotting = TRUE,
         bf.message = FALSE,
         title = expression(CaCO[3] ~ content))
#organic matter
om <- geochemistry1 %>%
  ggbetweenstats(x = layer, y = OM,
         type = "parametric",
         plot.type = "box",
         pairwise.comparisons = TRUE,
         pairwise.display = "significant",
         centrality.plotting = TRUE,
         bf.message = FALSE,
         title = "Organic matter (OM) content")
ggpubr::ggarrange(mg, om, toc_n, caco3, nrow=2, ncol=2)
#ggsave("differences.png", dpi=600, width=28, height=20, units = "cm")
```

References

Bertrand S, Tjallingii R, Kylander ME et al (2024) Inorganic geochemistry of lake sediments: a review of analytical techniques and guidelines for data interpretation. Earth Sci Rev 249:104639. https://doi.org/10.1016/j.earscirev.2023.104639

Boehmke B, Greenwell B (2020) Hands-on machine learning with R. CRC Press Taylor & Francis Group

Borcard D, Gillet F, Legendre P (2018) Numerical ecology with R, 2nd edn. Springer, Cham

Clarke KC (2013) What is the world's oldest map? Cartogr J 50:136–143. https://doi.org/10.1179/0008704113Z.00000000079

Daróczi G (2015) Matering data analysis with R. Packt Publishing Ltd., Birmingham

de Geer G (2002) Geochronology of the last 12,000 years. Int J Earth Sci 91:s100–s110. https://doi.org/10.1007/s00531-002-0287-6

Derrick B, Toher D, White P (2016) Why Welch's test is type I error robust. Quant Method Psychology 12:30–38. https://doi.org/10.20982/tqmp.12.1.p030

Friendly M, Sigal M (2014) Recent advances in visualizing multivariate linear models. Rev Colomb Estad 37:261–283. https://doi.org/10.15446/rce.v37n2spe.47934

Gower J, Gardner-Lubbe S, le Roux N (2011) Understanding biplots. Wiley, New York

Heywood I, Cornelius S, Carver S (1998) An introduction to geographical information systems. Addison Wesley Longman Limited, New York

Lesmeister C, Chinnamgari SK (2019) Advanced machine learning with R. Packt, Birmingham, Mumbai

Mauri A, Davis BAS, Collins PM, Kaplan JO (2015) The climate of Europe during the Holocene: a gridded pollen-based reconstruction and its multi-proxy evaluation. Quat Sci Rev 112:109–127. https://doi.org/10.1016/j.quascirev.2015.01.013

Oksanen J, Blanchet FG, Kindt R, et al (2018) "vegan" 2.3.3.—community ecology package

Patil I (2021) Visualizations with statistical details: the "ggstatsplot" approach. J Open Source Software 6:3167. https://doi.org/10.21105/joss.03167

Pawlik Ł, Okupny D, Kroh P et al (2022) Changing natural conditions and their impact on Mt Śnieżnica landscape, outer Western Carpathians—reconstruction of the Holocene environment based on geochemical and radiocarbon indices. Sci Total Environ 850:158066. https://doi.org/10.1016/j.scitotenv.2022.158066

Pearson K (1901) LIII. On lines and planes of closest fit to systems of points in space. Lond Edinb Dubl Phil Mag J Sci 2:559–572. https://doi.org/10.1080/14786440109462720

Sievert C (2020) Interactive web-based data visualization with R, plotly, and shiny. Chapman and Hall/CRC

Vu VQ, Friendly M (2024) ggbiplot: a grammar of graphics implementation of Biplots

Walker M (2005) Quaternary dating methods. John Wiley & Sons, New York

Wickham H (2007) Reshaping data with the reshape package. J Stat Softw 21:1–20

5 Climate Time Series

Simpson Desert, Australia, 2022, photo: Ł. Pawlik

Supplementary Information The online version contains supplementary material available at https://doi.org/10.1007/978-3-031-89673-6_5.

© The Author(s), under exclusive license to Springer Nature Switzerland AG 2025

Ł. Pawlik, *R Applications in Earth Sciences*, Springer Textbooks in Earth Sciences, Geography and Environment,
https://doi.org/10.1007/978-3-031-89673-6_5

In the 1940s, the stream of data had already become a flood…
Edwards (2010)

Times series store information on how a variable changes over time. This is a key property of many types of geoscience data. The dynamics of the data are evaluated in an expected order on the basis of fixed intervals, such as years, months, days, hours, minutes, or seconds. Since the first systematic meteorological observations, knowledge of climate characteristics has been used in investigations of climate evolution over time. With sets of historical observations and climate models, forecasting is possible but still extremely difficult. One iconic example illustrates the level of difficulty in weather/climate forecasting, and the sensitivity of models to small changes in input data is reflected by the so-called butterfly effect (Pascu 2012). The "butterfly" is the shape of Lorenz's strange attractor and is a common graphical explanation of chaos theory (El-Basha et al. 2016). The most common daily usage is a weather forecast that can potentially be valid for the next several hours but is never 100% correct. Weather reflects just a slice of the greater atmosphere dynamics at a given moment and must be forecasted considering climate conditions. To assess climate conditions and types correctly, at least 30 years of observations are needed. This period is the interval recommended by the World Meteorological Organization (WMO). According to the WMO, climate reflects the average weather conditions at a particular location over a long period, ranging from months to thousands or millions of years (https://wmo.int/topics/climate). High-quality climate data are necessary if precise climate predictions are expected. This task normally involves massive databases, long-term time series from various sources, complex algorithms, and high-speed computer clusters.

The collection of observations for time series construction is a difficult task and depends on many factors. As a result, what we obtain as a final product for statistical analyses and exploration does not, in principle, contain precise information. A data analyst must understand that datasets are rarely complete or 100% correct. The first constraint is related to our ability to measure everything everywhere. Obviously, this is impossible. Traditionally, climate properties are measured at climate stations with limited spatial density, and measurement intervals are normally at the hourly scale (Fig. 5.1). Measurements at shorter intervals are possible (even in seconds or minutes), but in such a situation, we may face problems with a large data volume, as such datasets are not easy to process with computers or individual machines with limited operational storage. Finally, for some extremely remote and inhospitable places, there are no meteorological observations. Notably, the "white spots" problem on the world map

Fig. 5.1 Automatic climate station installed on Mt. Babia Góra, S. Poland. The station measures temperature, humidity, wind speed, and direction at 10-minute intervals. Photo: Łukasz Pawlik

of direct meteorological observations has been eliminated by the usage of satellite products, which are now commonplace.

Time series, which are commonly analyzed in fields such as climatology or hydrology, can be explored and decomposed to assess their main components, such as trends, cycles, and seasonality (Hyndman and Athanasopoulos 2021). A trend is a component that develops over long-term intervals. If we analyze, for instance, daily, monthly, or quarterly data, a trend is normally visible (detectable) over a 3–4-year period. A cycle is a time series property that is shorter than a trend and longer than a seasonal component, the latter being more intuitive when climate seasons are considered.

5.1 Description of the Data and Project Outline

For this project, two climate stations are selected. One station is located in Muzaffarabad city in Pakistan, and the other is located in Chandigarh city in India. We used the data archive of the GSOD (provided by the National Centers for Environmental Information, https://www.ncei.noaa.gov/access/metadata/landing-page/bin/iso?id=gov.noaa.ncdc:C00516). Access to this database is possible using the **GSODR** package (Sparks et al. 2024). Without the exact geographical coordinates of a station for a selected locality (cities, as in this case), we can define them using, for example, Google Maps. With information on latitude and longitude, the ***nearest_stations()*** function is used for searching climate stations with data records for a given locality. The data used in this project are of daily resolution (daily mean values). For Muzaffarabad, defining the arguments LAT, LON, and distance at 10 km, we received information on two stations with observations covering different periods, 1973–2024 and 2005–2020. Only one station was found in Chandigarh, with observations covering the period 2006–2024. The geographic position of the climate stations can be easily displayed on an interactive map using the **leaflet** package (Cheng et al. 2023) (Fig. 5.2). In this initial step, we obtained the following feedback information: (1) different numbers of climate stations in the area, (2) different observation periods at each station, and (3) large variability

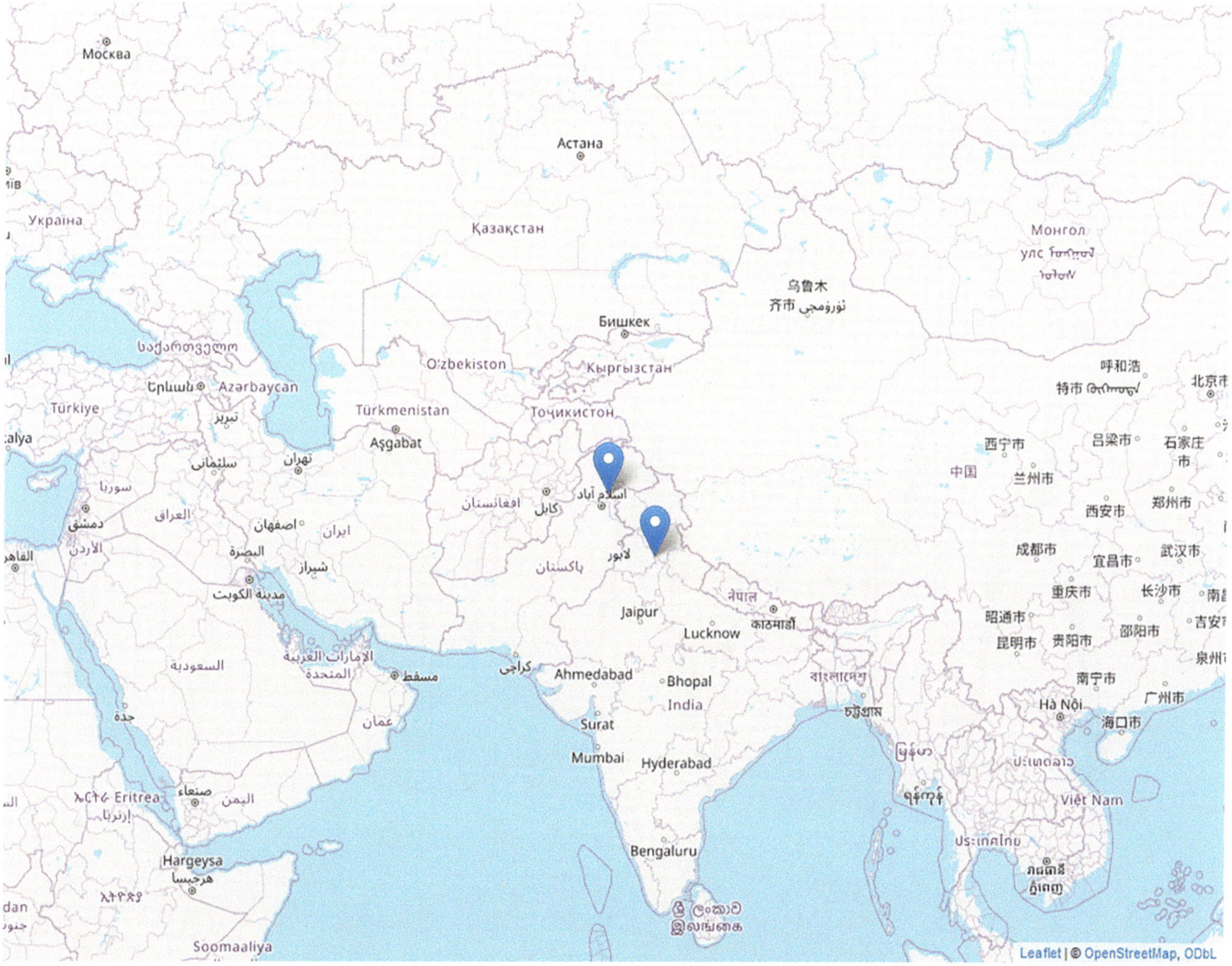

Fig. 5.2 Geographic position of two selected stations: Muzaffarabad in Pakistan and Chandigarh in India

in the data quality (e.g., proportion of missing observations). The first station in Muzzafarabad contains only eight observations, and the dataset for the second station is largely incomplete (Fig. 5.3). After downloading and testing the data for the selected station in India, Chandigarh, the time series was also found to be incomplete. However, when inspecting the data, NAs, indicating missing observations, were not present. The dates with missing observations were omitted and not included in the dataset. These issues need to be fixed before time series construction. To build a time series (*ts*) object in R and impute the data, a complete dataset is needed, with a full set of dates for the analyzed period. First, a sequence of dates was created, after which the one-column data were combined with the table downloaded from the GSOD archive (Fig. 5.4). The proportion of missing observations (Fig. 5.5) was determined using the ***missmap()*** function from the **Amelia** package (Honaker et al. 2011).

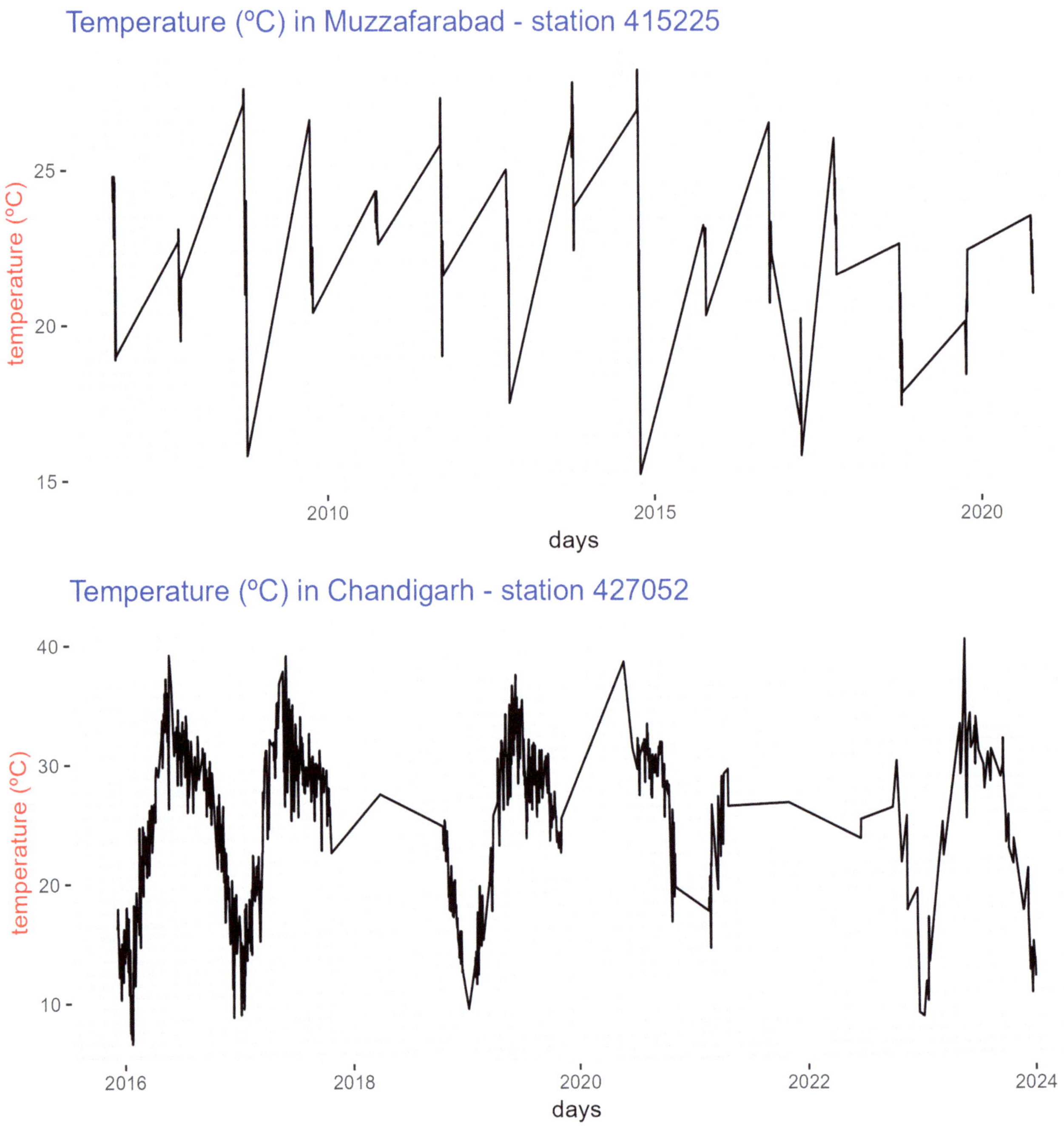

Fig. 5.3 Temperature time series for selected climate stations in Pakistan and India

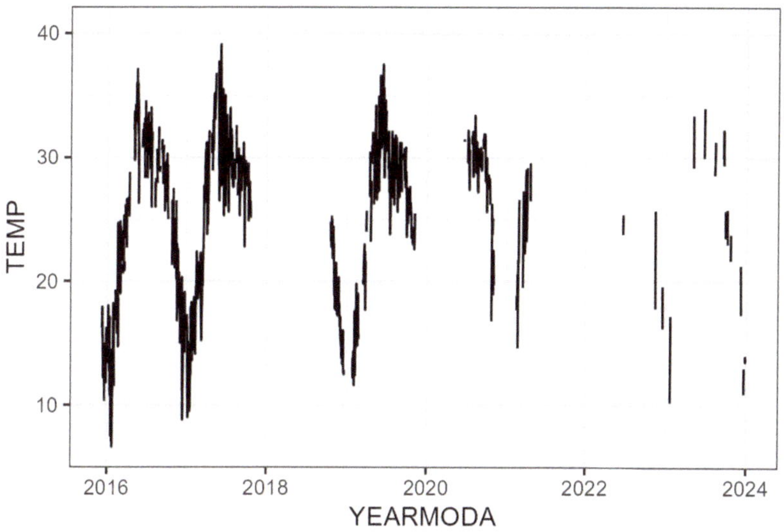

Fig. 5.4 A plot for the mean daily temperature in Chandigarh with gaps indicating missing values

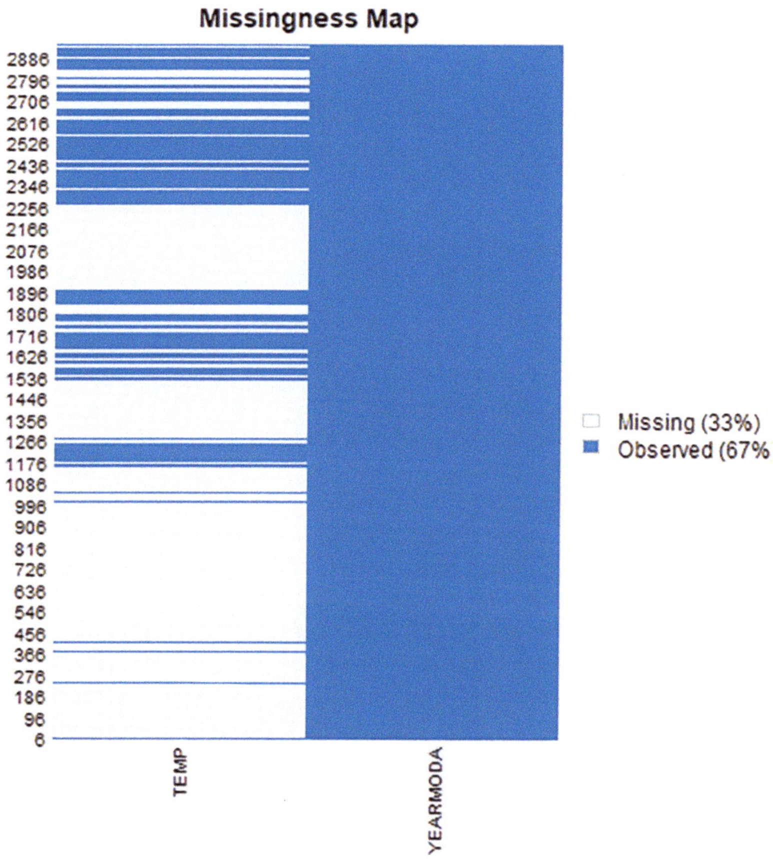

Fig. 5.5 A map of missing values for the mean daily temperature observations in Chandigarh

```
library(leaflet) #dynamic and interactive maps
library(mapview) #interactive viewing of
spatial data
#Two selected stations on the World map.
cs <- leaflet() %>% setView(lng = 74, lat = 35,
zoom=4) %>%
 addTiles() %>% addMarkers(73.46787147731179,
34.357776459608, label = "Muzaffarabad") %>%
 addMarkers(76.77709324290086,
30.7281149621589, label = "Chandigarh")
cs
#Save the map as an image.
#mapshot(cs, file = "climate_stations.png")
pacman::p_load("GSODR", "ggplot2",
"rnaturalearth", "dplyr", "rmarkdown",
```

```
"ggpubr", "leaflet", "lubridate", "mapview",
"forecast", "imputeTS",
"gt", "magick")
library(GSODR)
library(ggplot2)
library(rnaturalearth)
library(dplyr)
#We will try to find the nearest station to
Muzaffarabad and Chandigarh.
muzaffarabad <- nearest_stations(LAT =
34.357776459608, LON = 73.46787147731179,
 distance=10)
chandigarh <- nearest_stations(LAT =
30.7281149621589, LON = 76.77709324290086,
 distance=10)
#Display this information with the rmarkdown
package.
library(rmarkdown) #converts R Markdown
documents into a variety of formats
station_info <- rbind(muzaffarabad, chandigarh)
paged_table(station_info)
#or
head(station_info)
#With information from the table, we can set a
correct interval for data download.
#The first station (415320) contains only eight
observations, so we will focus on the second
station.
#Download the data.
muzzaffarabad1 <- get_GSOD(years = 2006:2020,
station = "415225-99999")
#Data for the station in India.
chandigarh1 <- get_GSOD(years = 2006:2023,
station = "427052-99999")
#With this dataset, we get more observations,
but still, it is largely incomplete.
#Note that all units are converted to
International System of Units (as written in a
description of the get_GSODR function).
#?get_GSOD #call the function description
#How has the temperature changed in
Muzzafarabad over the past decades?
plot1 <- ggplot()+
  geom_line(data = muzzafarabad1, mapping =
  aes(x=YEARMODA, y=TEMP))+
  ggtitle("Temperature (°C) in Muzzafarabad -
  station 415225")+
  ylab("temperature (°C)")+
  xlab("days")+
  theme
 (plot.title = element_text(color = "blue"),
        axis.title.y = element_text(color =
        "red"))
 #With this dataset, we get largely incomplete
 information.
plot2 <- ggplot()+
  geom_line(data = chandigarh1, mapping =
  aes(x=YEARMODA, y=TEMP))+
  ggtitle("Temperature (°C) in Chandigarh -
  station 427052")+
  ylab("temperature (°C)")+
  xlab("days")+
  theme
(plot.title = element_text(color = "blue"),
axis.title.y = element_text(color = "red"))
#Note that YEARMODA variable is in "Date"
format. ggplot2 will recognize the date format
and use dates correctly to construct the time
series.
#Arrange two plots in one figure.
library(ggpubr)
ggarrange(plot1, plot2, nrow = 2)
#Save the plot.
#ggsave("cliamate_stations.png")
#Adding missing dates to the time series.
library(lubridate) #functions to work with
date-times and time-spans
#We first define the sequence of dates for which
values will be interpolated. The sequence is
built following the information from the first
cells of the YEAR, MONTH, and DAY columns,
converted into the date format using the ymd()
function, and the information on the dataset
length we got after using the length()
function.
#The result is a table with calendar dates
only.
empty_df <- data.frame(YEARMODA = seq(ymd(paste
(chandigarh1[1,YEAR], chandigarh1[1,MONTH],
chandigarh1[1,DAY], sep="-")), ymd(paste(chandi
garh1[length(chandigarh1$STNID),YEAR], chandiga
rh1[length(chandigarh1$STNID),MONTH], chandigar
h1[length(chandigarh1$STNID),DAY], sep="-")),
by = 'days'))
#Now, we can combine it with the dataset but
only select the information on temperature.
chandigarh1a <- left_join(empty_df,
chandigarh1, by = "YEARMODA") %>%
 dplyr::select(YEARMODA, TEMP)
#How does it look with NA values?
ggplot()+
 geom_line(chandigarh1a, mapping =
aes(x=YEARMODA, y=TEMP))+
 theme_bw()
#Save the plot.
#ggsave("ts_chandigarhNA.png", width = 12,
height = 8, units = "cm")
#The dataset can be inspected using the View()
function or by clicking the dataset name in the
Environment panel in RStudio.
```

```
#By using the summary() function, we get
information about 1949 NA values.
summary(chandigarh1a) #is this a good way?
#1949 NA values are over 66% of the
observations.
length(chandigarh1a[chandigarh1a$TEMP ==
"NA",2])/length(chandigarh1a$YEARMODA)*100
#Another way to check the data completeness.
#png("missmap.png") #opening the graphic device
to save the image
library(Amelia) #package for missing data
missmap(chandigarh1a)
#dev.off()#shutting down the png graphic device
#The plot shows 33% of missing observations,
but it takes into account the "YEARMODA"
column.
length(chandigarh1a[chandigarh1a$TEMP ==
"NA",2])/(2*length(chandigarh1a$YEARMODA))*100
#This code is also an example of how, in the R
function, outcomes representing specific values
can be mixed with numbers during calculations.
```

5.2 R in Practice: Climate Time Series Evaluation and Exploration

Although it is not recommended for datasets with large gaps (Lepot et al. 2017), the next step is data imputation using selected interpolation methods. The selected time series contains information on the mean daily surface temperature and features a strong seasonal component. This component is used for data imputation with the ***na_interp()*** function from the **forecast** package (Hyndman et al. 2024). The upper panel in Fig. 5.6 shows the effect of this operation in comparison with the result of the ***na_interpolation()*** function from the **imputTS** package (Moritz and Bartz-Beielstein 2017). For the seasonal series, the ***na_interp()*** function first computes the STL decomposition (seasonal and trend decomposition using LOESS (locally estimated scatterplot smoothing; Cleveland et al. 1990)). When the data are seasonally adjusted (i.e., the seasonal component is removed), linear interpolation is applied, and the seasonal component is added back to the series. For the data used in this example,

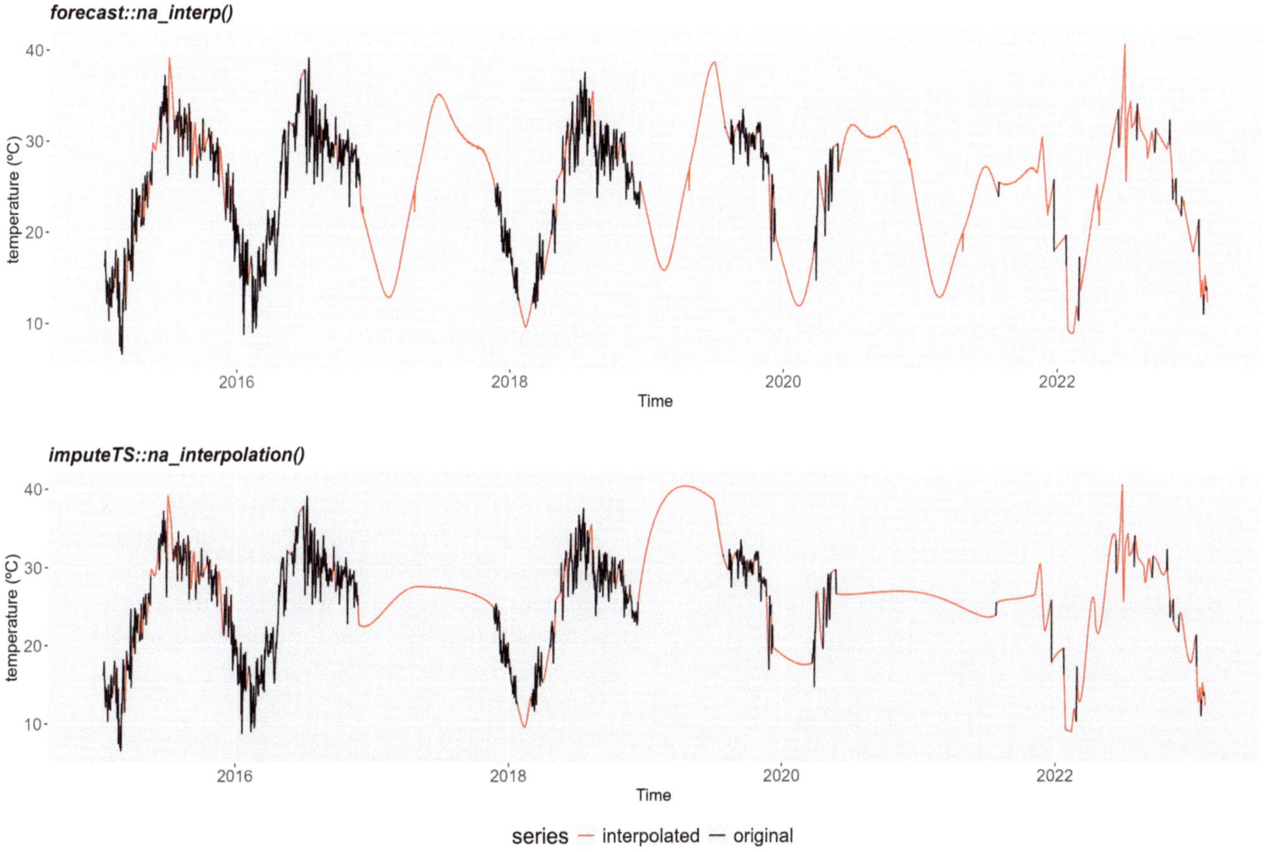

Fig. 5.6 Time series of daily temperature observations from Chandigarh, with gaps filled by applying two interpolation methods: *na.interp()* (upper panel) and *na.interpolation()* (lower panel)

the ***na_interpolation()*** function yields the best results with the argument *option="stine"*, which is the Stineman interpolation (Stineman 1980). The method was implemented in the **stinepack** library (Johannesson et al. 2024).

Stineman highlighted the following properties of the method:

1. If the values of the ordinates of the specified points change monotonically and the slopes of the line segments joining the points change monotonically, then the interpolated curve and its slope will change monotonically.
2. If the slopes of the line segments joining the specified points change monotonically, then the slope of the interpolated curve will change monotonically.
3. Suppose that the conditions in (1) or (2) are satisfied by a set of points but that a small change in the ordinate or slope at one of the points will result in conditions (1) or (2) being no longer satisfied. This small change in the ordinate or slope at a point will cause no more than a small change in the interpolated curve (Stineman 1980).

The ***na_interp()*** function yields visually better results in the analyzed case than do the other functions (Fig. 5.6). The ***na_interpolation()*** function smooths the parts of the curve associated with the warm and cold seasons; however, it works well for periods with few data gaps.

```
#Now, we can transform the data into a ts
object containing the mean daily temperature
since 2005.
chandigarh1a_ts <- ts(chandigarh1a$TEMP, start
= c(2015,12,08), frequency = 365.25)
#Time series interpolation with two functions
applied to the same dataset.
library(forecast) #methods and tools for
displaying and analyzing univariate time series
forecasts
chandigarh1a_int <- na.interp(chandigarh1a_ts)
library(imputeTS) #time series missing value
imputation
chandigarh1a_int1 <- na.
interpolation(chandigarh1a_ts, option =
"stine")
#Plotting the time series.
plot1_int <- autoplot(chandigarh1a_int, series
= "interpolated")+
 autolayer(chandigarh1a_ts, series =
"original")+
 scale_color_manual(values = c("interpolated" =
"red", "original"="black"))+
 ylab("temperature (°C)")+
 labs(subtitle = "forecast::na_interp()")+
 theme
(legend.title = element_text(size=18),
 legend.text = element_text(size=16),
 axis.text.x = element_text(size=14),
 axis.text.y = element_text(size=14),
 axis.title.x = element_text(size=14),
        axis.title.y = element_text(size=14),
        legend.position = "none",
        plot.subtitle = element_text(size = 16,
face = "bold.italic"))
#Plotting the time series.
plot2_int <- autoplot(chandigarh1a_int1, series
= "interpolated")+
  autolayer(chandigarh1a_ts, series =
"original")+
  scale_color_manual(values = c("interpolated"
= "red", "original"="black"))+
  ylab("temperature (°C)")+

labs(subtitle
= "imputeTS::na_interpolation()")+
  theme
(legend.title = element_text(size=18),
        legend.text = element_text(size=16),
        axis.text.x = element_text(size=14),
        axis.text.y = element_text(size=14),
        axis.title.x = element_text(size=14),
        axis.title.y = element_text(size=14),
        legend.position = "bottom",
        plot.subtitle = element_text(size = 16,
face = "bold.italic"))
#Arranging two plots.
ggarrange(plot1_int, plot2_int, nrow = 2)
#Saving the plot.
#ggsave("interpolated_temp_two_methods.png",
width = 15, height = 10, units = "cm")
```

The time series features a strong seasonal component in the annual pattern of daily surface temperature (Fig. 5.7). This is a natural pattern in which the Earth's surface is exposed to solar radiation, which causes temperature fluctuations during the warm and cold seasons. However, importantly in this specific example, the time series was used after interpolation. The data quality, i.e., the unavailability of a complete dataset, prevents the formulation of strong conclusions regarding a data trend. The remainder, which is information beyond the seasonal and trend components, indicates that there is not a large portion of unexplained data variability.

```
#Decomposition of the time series using the STL
method (seasonal, trend, and irregular
components using loess). The time series was
modified by filling the gaps during the
interpolation procedure.
chandigarh1a_int_stl <- mstl(chandigarh1a_int,
lambda = "auto")
#png("stl_decomposition.png")#opening the
graphic device for saving the image
```

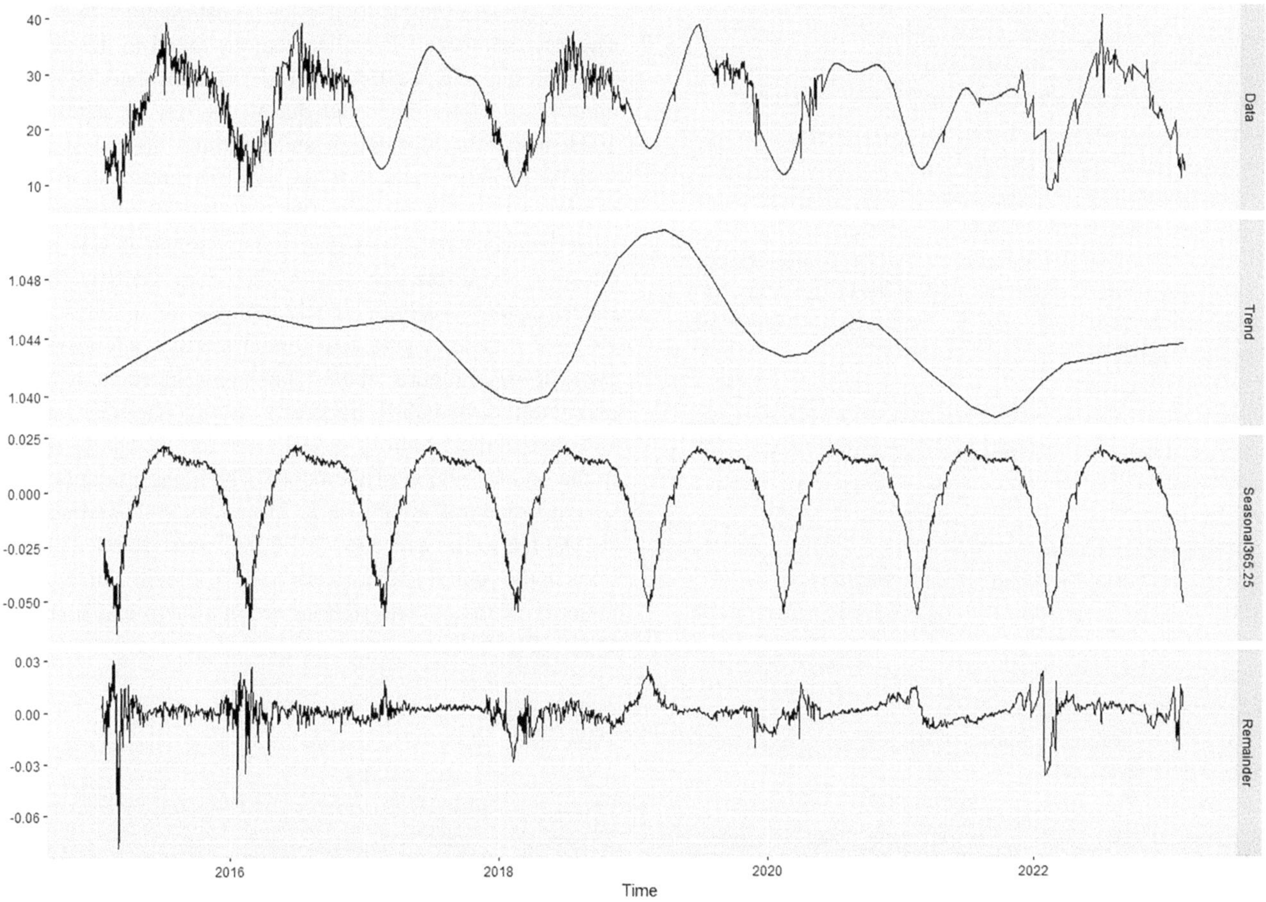

Fig. 5.7 Results of the STL decomposition of the time series containing observations of the mean daily temperature in Chandigarh, India

```
autoplot(chandigarh1a_int_stl)
#dev.off()#shutting down the png graphic device
#We will now find only 5% of days with the
highest temperatures.
#This task can be done by using the quantile()
function.
#For Chandigarh (we use data before
interpolation).
temp_max_chandigarh <-
quantile(chandigarh1a$TEMP, 0.95, na.rm = TRUE)
temp_max_chandigarh#what is the temperature
value for the 95 percentile?
#Now, we need to filter the dataset.
chandigarh_data95 <- chandigarh1a %>%
 dplyr::filter(TEMP > temp_max_chandigarh)
#How many observations are left?
str(chandigarh_data95)
#Which day was the hottest day according to
this data?
chandigarh1a |> dplyr::filter(TEMP == max(TEMP,
na.rm = TRUE))
#Or it can be found out in this way.
subset(chandigarh1a, TEMP == max(TEMP, na.rm =
TRUE))
#Or in this way:
chandigarh1a[which.max(chandigarh1a$TEMP),]
#Which year has the highest number of days with
a temperature over the 95 percentile?
#Note that the dataset before interpolation is
used here.
temp_frequency_chandigarh1 <- chandigarh1 %>%
 dplyr::group_by(YEAR) %>%
 dplyr::filter(TEMP > temp_max_chandigarh) %>%
 dplyr::summarize(n_days = length(TEMP), max_
temp95 = max(TEMP, na.rm = TRUE))
temp_frequency_chandigarh1
#We now change the column names.
colnames(temp_frequency_chandigarh1) <-
c("year", "days", "max_temp")
#The table can be saved as an image using the
gt library.
library(gt) #package supporting building
display tables
#gtsave(gt(temp_frequency_chandigarh1), "temp_
max_table.png")
#We can plot this data.
ggplot()+
```

```
 geom_col(data = temp_frequency_chandigarh1,
mapping = aes(x=year, y=days)) +
 ylab("number of days")+
 ggtitle(paste0("Number of days with the
temperature over ",temp_max_chandigarh, " (95
percentile)"))+
 labs(subtitle = "in Chandigarh (2015-2023)")
#ggsave("bar_plot_temperature.png")
#Using the magick package, we combine the bar
chart and the table of the annual number of
days with the mean daily temperature over the
95 percentile.
library(magick)
#Read the images saved in the previous steps.
table1_image <- image_read("https://raw.github.
com/Lukpawlik/SN_figures/main/temp_max_table.
png") |>
 image_scale("250x")
bar_plot_temp <- image_read("https://raw.
github.com/Lukpawlik/SN_figures/main/bar_plot_
temperature.png") |>
 image_scale("600x")
#Combine two images.
image2 <- image_composite(bar_plot_temp,
table1_image, offset = "+340+70")
image 2
#image_write(image2, "combined_temp_images.
png")
```

For long time series with observations at a high temporal resolution (hours or days), we might be interested in calculating some general characteristics, such as the 95th percentile and the maximum daily temperature recorded during a year. The 95th percentile represents a threshold value that reduces the dataset to the top 5% of observations with the highest values in the dataset. The threshold value for the 95th percentile can be calculated using the ***quantile()*** function from the **stats** package. The daily temperature in Chandigarh is 33.8 °C, and this value is used to filter the time series via the ***dplyr::filter()*** function. When attempting to find the maximum annual temperature, a very convenient approach involves the use of the ***dplyr::group_by()*** and ***dplyr::summarize()*** functions. These functions are used in a code chain via the %>% pipe operator provided in the **magrittr** package (Bache and Wickham 2022) (Sect. 2.2.1). They are applied one after another from the top to the bottom of the code chunk. Figure 5.8 displays the results of the ***summarize()*** function in plot and table form. In 2017, there were 19 days with daily temperatures above the 95th percentile threshold, and the maximum daily temperature was recorded in 2023. The table was saved with the ***gtsave()*** function from the **gt** package (Iannone et al. 2024), imported and then inserted into the ggplot2-based plot with the ***image_composite()*** from the **magick** package (Ooms 2024).

The second example focuses on climate stations in Brazil. The total number of Brazilian stations found in the GSOD database was 479 in 2023. The task is to calculate the maximum temperature measured during the day at stations in 2023, map this information and highlight the ten stations with the highest values. In R, the user frequently imports and exports data in tabular form (the data frame and tibble, the latter being an improved data frame format; it was introduced in the **tibble** package (Müller et al. 2018)). This is easy to display in RStudio or RMarkdown, but in some cases, we need a table in graphical format that is ready to use in publications or reports. Table 5.1 contains information on the maximum temperature recorded during the day at 10 Brazilian climate stations in 2023 (the full list can be found in the online version of the book). The highest temperature was recorded in Parnaiba on 27 December 2023 and was 45 °C. All ten values exceeded 40 °C and were recorded at stations in the central and northern parts of Brazil. Notably, the majority of the climate stations are in eastern and southern Brazil (Fig. 5.9).

```
pacman::p_load("rgeoboundaries", "ggrepel")
#The next step is to download the selected
country and year data.
brazil <- get_GSOD(years=2023, country =
"Brazil")
#Station names.
unique(brazil$NAME)
#The data for 2023.
max_temp_brazil <- brazil %>%
  group_by(NAME, LATITUDE, LONGITUDE) %>%
  mutate(max_temp = max(MAX)) %>%
  filter(MAX == max_temp)
max_temp_brazil
#Arranging the table.
max_temp_brazil1 <- max_temp_brazil %>%
  dplyr::arrange(desc(max_temp)) %>%
  select(NAME, LATITUDE, LONGITUDE, max_temp,
YEARMODA) %>%
  as.data.frame()
max_temp_brazil1
max_temp_brazil2 <- max_temp_brazil1[1:10,] %>%
  gt() %>%
  cols_width(everything() ~ px(150)) %>%
  data_color(columns = max_temp,
             palette = "viridis")
#Saving the table.
#gtsave(max_temp_brazil2, "brazil_max_temp_
table.png")
#We can show this data on the map.
library(rgeoboundaries) #provides access to the
geoBoundaries international boundary database
https://www.geoboundaries.org
brazil_map <- gb_adm0("Brazil")
```

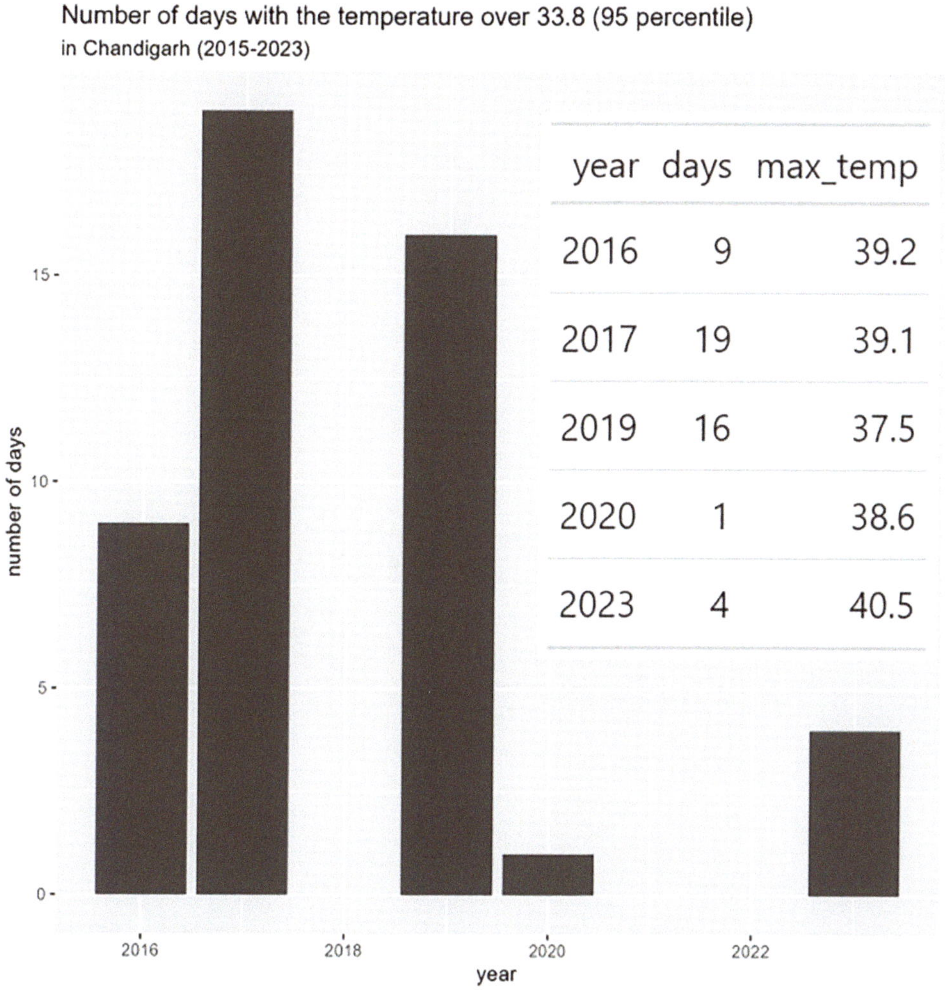

Fig. 5.8 Number of days per year with temperatures above the 95th percentile in Chandigarh, India, and the annual maximum temperature

Table 5.1 Daily maximum temperature and 10 highest values measured at climate stations in Brazil in 2023. Data source: GSOD. The values in the max_temp column are colored using the ***data_color()*** function from the **gt** package

NAME	LATITUDE	LONGITUDE	max_temp	YEARMODA
PARNAIBA	-3.067	-41.783	45.0	2023-12-27
ARAGARCAS	-15.900	-52.250	44.3	2023-10-19
RONDON DO PARA	-4.783	-48.067	43.8	2023-10-20
SAO ROMAO	-16.367	-45.117	43.5	2023-09-26
PORTO MURTINHO	-21.700	-57.550	43.4	2023-10-17
PORTO MURTINHO	-21.700	-57.550	43.4	2023-11-16
TURIACU	-1.667	-45.367	43.3	2023-08-20
CUIABA	-15.617	-56.100	43.3	2023-10-19
CORUMBA	-19.000	-57.633	43.3	2023-11-14
RONDONOPOLIS	-16.450	-54.567	42.8	2023-10-19

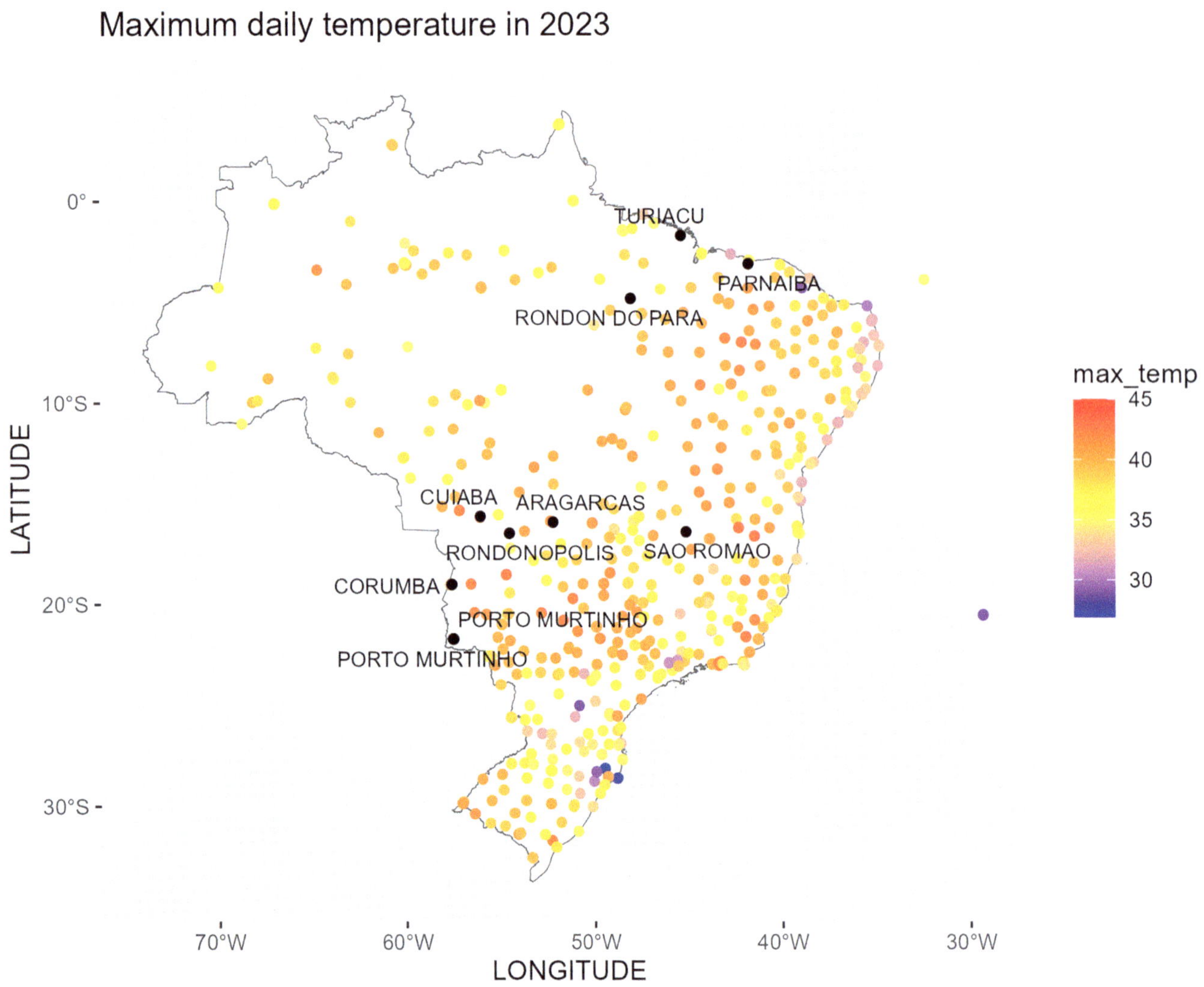

Fig. 5.9 Maximum daily temperature in Brazil in 2023. The map was constructed using the **ggplot2** package

```
#Plotting.
library(ggrepel) #provides text and label geoms
for 'ggplot2' that help to avoid overlapping
text labels
ggplot()+
  geom_sf(data = brazil_map, fill="white")+
  geom_point(data = max_temp_brazil,
             mapping = aes(x = LONGITUDE, y =
             LATITUDE,
                          color = max_temp))+
  geom_point(max_temp_brazil1[1:10,], mapping =
  aes(x = LONGITUDE, y = LATITUDE))+
  geom_text_repel(max_temp_brazil1[1:10,],
  mapping = aes(x = LONGITUDE, y = LATITUDE,
  label = NAME), size=3)+
  scale_color_gradientn(colors = c("blue",
  "yellow", "red"))+
  ggtitle("Maximum daily temperature in 2023")
#Saving the map.
#ggsave("brazil_max_temp1.png")
```

References

Bache S, Wickham H (2022) magrittr: a forward-pipe operator for R

Cheng J, Schloerke B, Karambelkar B, Xie Y (2023) Leaflet: create interactive web maps with the JavaScript “leaflet” library. R package version 2.2.0

Cleveland RB, Cleveland WS, McRae JE, Terpenning (1990) STL: a seasonal-ternd decomposition procedure based on loess. J Off Stat 6:3–33

Edwards PN (2010) A vast machine: computer models, climate data, and the politics of global warming. The MIT Press, Cambridge

El-Basha O, El-Shahat A, Fayed H (2016) Chaos theory and Lorenz attractors. Sohag J Sci 1:7–12

Honaker J, King G, Blackwell M (2011) Amelia II: a program for missing data. J Stat Softw 45:1–47

Hyndman RJ, Athanasopoulos G (2021) Forecasting: principles and practice, 3rd edn. OTexts, Melbourne

Hyndman RJ, Anthanasopoulos G, Bergmeir C, et al (2024) Forecast: forecasting functions for time series and linear models

Iannone R, Cheng J, Schloerke B, et al (2024) gt: easily create presentation-ready display tables

Johannesson T, Bjornsson H, Icelandic Met. Office, Grothendieck G (2024) stinepack: Stineman, a Consistently Well Behaved Method of Interpolation

Lepot M, Aubin J-B, Clemens FHLR (2017) Interpolation in time series: an introductive overview of existing methods, their performance criteria and uncertainty assessment. Water 9:796. https://doi.org/10.3390/w9100796

Moritz S, Bartz-Beielstein T (2017) imputeTS: time series missing value imputation in R. R J 9:207–218. https://doi.org/10.32614/RJ-2017-009

Müller RD, Cannon J, Qin X et al (2018) GPlates: building a virtual earth through deep time. Geochem Geophys Geosyst 19:2243–2261. https://doi.org/10.1029/2018GC007584

Ooms J (2024) magick: advanced graphics and image-processing in R

Pascu RI (2012) Weather prediction and the butterfly effect. Sci Papers 1:109–113

Sparks A, Hengl T, Nelson A (2024) GSODR: global summary daily weather data in R

Stineman RW (1980) A consistently well behaved method of interpolation. Creative Computing 6:54–57

6 Geomorphic Data and Geomorphometry Analyses

Mammoth Hot Springs, Yellowstone National Park, U.S.A., 2019, photo: Ł. Pawlik

General geomorphometry provides an extremely useful basis for geomorphology, if only in suggesting the important spatial variations and structures that this science is to explain.
Evans (1972)

Supplementary Information The online version contains supplementary material available at https://doi.org/10.1007/978-3-031-89673-6_6.

Geomorphology is a scientific discipline focused on Earth's landscapes and landforms and the processes of their formation (Bloom 1969; Pitty 1971). Moreover, it is closely related to physical geography and geology. Traditionally, geomorphologists have studied landforms, but with advancements in remote sensing techniques, the range of potential study objects has spread to extraterrestrial objects and ocean

© The Author(s), under exclusive license to Springer Nature Switzerland AG 2025

Ł. Pawlik, *R Applications in Earth Sciences*, Springer Textbooks in Earth Sciences, Geography and Environment,
https://doi.org/10.1007/978-3-031-89673-6_6

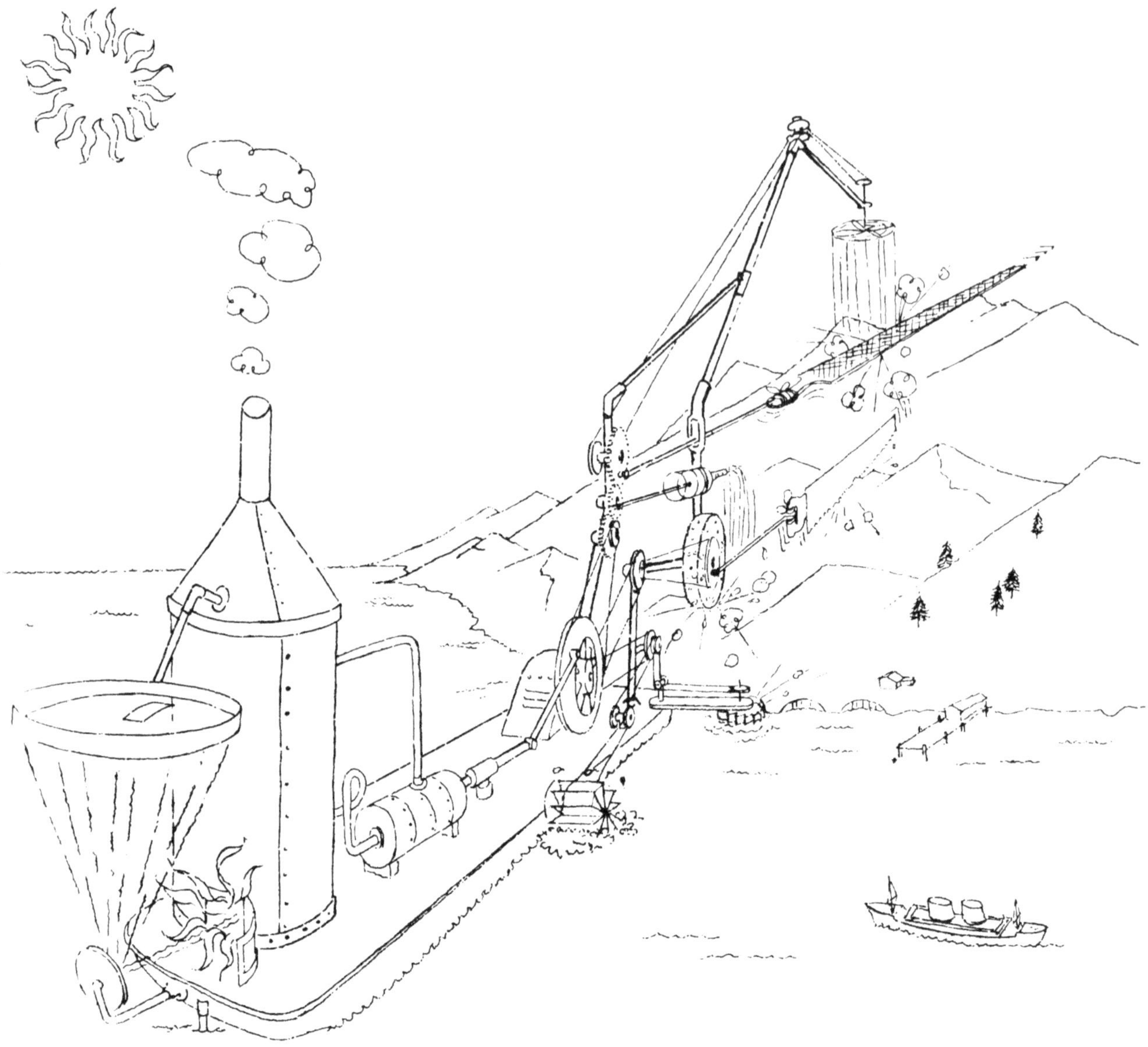

Fig. 6.1 The external forces acting upon the mountain landscape. A sketch from the Surface of the Earth book (Bloom 1969). The original figure caption is "*The 'geomorphology machine' (after the style of Rube Goldberg)*"

bottoms (Wilson et al. 2007). Currently, data collection even includes ground images captures by rovers, such as the Curiosity rover (Diniega et al. 2021). The subject of geomorphological interest was expressed as "*the events between the creation and the final destruction*" of landforms (Huggett 2007). This concept is well illustrated in Fig. 6.1, which shows, in a metaphorical way, the destruction of a mountain range by external factors (Bloom 1969). These factors include weathering processes, erosion, and abrasion. In addition, some scientists view geomorphology as a discipline that explains "*how landforms and landscapes respond to variable boundary conditions and disturbances*" and "*how geomorphic systems coevolve with climate, ecosystems, soils, and other environmental systems*" (Phillips 2009). It represents a more systematic approach to physical geography that started in the 1960s (Chorley and Kennedy 1971).

Geomorphometry evolved from mathematics, geography, and geomorphology as a science oriented to quantitative land-surface analysis (Pike et al. 2008). It has been characterized in the following ways:

1. A modern, analytical-cartographic approach for representing bare-Earth topography via computer manipulation of terrain height.
2. It involves discrete surface features (landforms) and continuous land surfaces (Evans 1972).

Geomorphometry provides an important set of tools and analytical approaches for geomorphological analysis (Pawlik et al. 2024), digital soil mapping (Temme et al. 2022), and ecological assessment (Ironside et al. 2018). Geomorphometric analyses are mostly based on DEMs and bare-earth raster

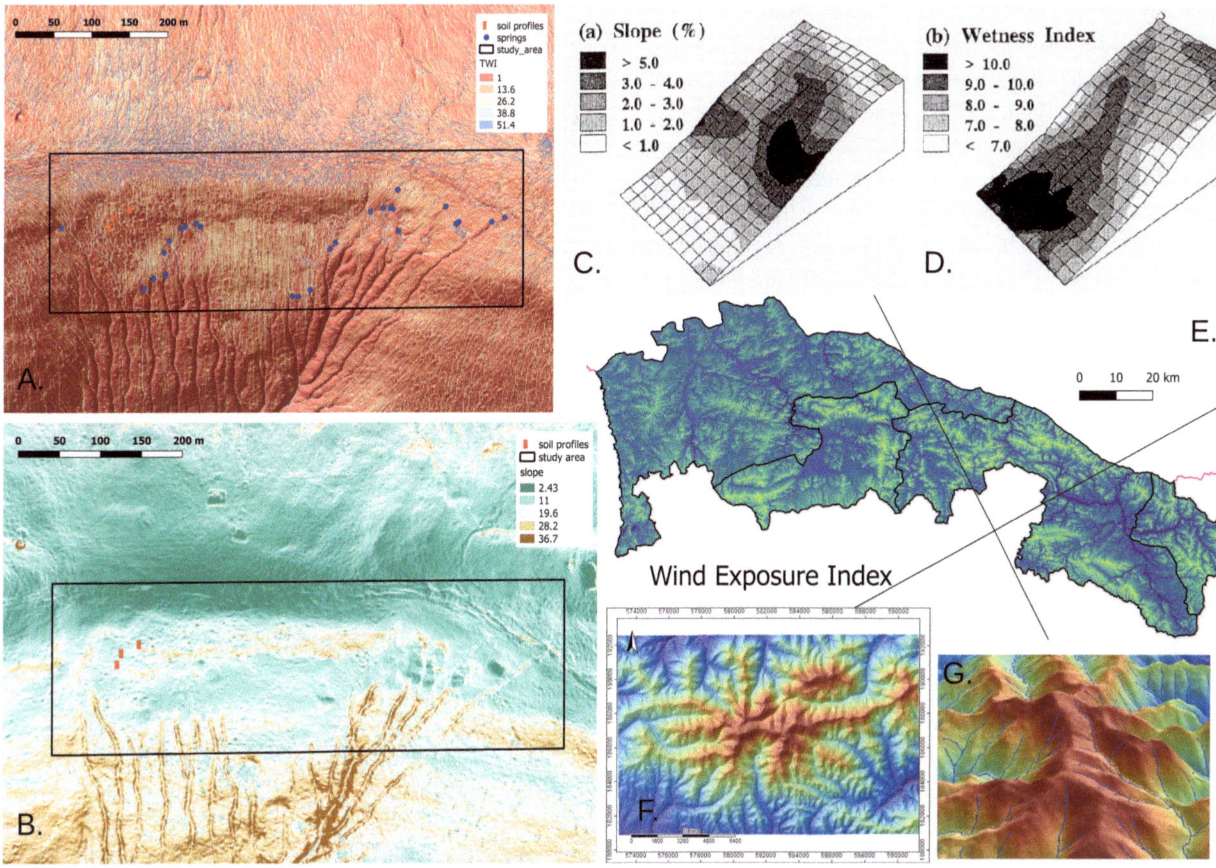

Fig. 6.2 Examples of various geomorphometric parameters: (**a**) topographic wetness index (TWI) calculated in SAGA GIS for the central part of Gorce National Park, Poland (map prepared in QGIS); (**b**) slope (in degrees) calculated for the same area as in A (Pawlik et al. 2019); (**c**, **d**) slope (in %) and wetness index (Moore et al. 1993); (**e**) wind exposure index (WEI) calculated for the Polish Sudety Mountains, SW Poland (Pawlik and Harrison 2022); (**f**) elevation raster model for the Gorce Mountains region in Poland; and (**g**) 3D model of a mountain ridge with a first-order stream network

grids, which are constructed using various data sources and techniques, such as topographical map vectorization, satellite images from radar systems, and LiDAR-based (Light Detection and Ranging) scanning. The most common geomorphometric derivatives calculated from DEMs are the surface slope, aspect (terrain orientation) (Fig. 6.2), topographic position index (TPI), surface roughness, terrain ruggedness index (TRI), and flow direction (Olaya 2008; Xiong et al. 2022). These, among other parameters, are controlling factors of many surface processes (Moore et al. 1993; Lee and Clarke 2005). The topographic slope is the first derivative of elevation, which is given as the change in elevation over distance (Lee and Clarke 2005). The surface slope parameter is critical for evaluating, modeling, and predicting hillslope stability, as well as assessing geohazards, such as landslides, debris flows, and rock falls (Pawlik et al. 2023). The aspect parameter is critical in microclimate studies and vegetation modeling, as some plants grow best on south-facing slopes; moreover, tree damage can be greatest on hillslopes exposed to the strongest winds (Dyderski and Pawlik 2020; Pawlik and Harrison 2022). The TPI, TRI, and roughness represent local differences in elevation, which can be important predictors of plant redistribution and useful in soil erosion modeling. The TRI is the mean of the absolute difference between the value of a cell and its eight surrounding cells (Riley et al. 1999). The TPI is a measure of relative position and reflects whether an area is a local topographic high (positive values) or low (negative values) (Ilich et al. 2023). It is the difference between the elevation at a central point and the average elevation around this point within a predefined radius (Al-Sababhah 2023).

6.1 Description of the Data and Project Outline

Elevation data with global coverage can be obtained from several geodatabases and missions. One of the most popular DEMs was generated with data from the Shuttle Radar Topography Mission (SRTM), with an original resolution of 30 m (Rodriguez et al. 2005). These data can be conveniently imported into a project for the selected locality, defined, for example, by a polygon with a state border, using the ***elev()*** function from the **climenv** package (Tsakalos et al. 2023). Although SRTM data have a relatively low spatial resolution, they can be used to demonstrate some of the most popular concepts applied in geomorphometry. For data selection and download operations, we use the **rgeoboundaries** package (Dicko 2024) and import a vector model with a border of Vietnam. The Vietnam border, a spatial object, is used to constrain the import of a raster with SRTM elevation data for the selected area. *SpatRaster*, an object class from the **terra** package, is created in this process (Hijmans 2023). The polygon with the Vietnam border is also used to mask the data and reduce the raster to the area within the polygon (Figs. 6.3, 6.4, 6.5, 6.6, and 6.7).

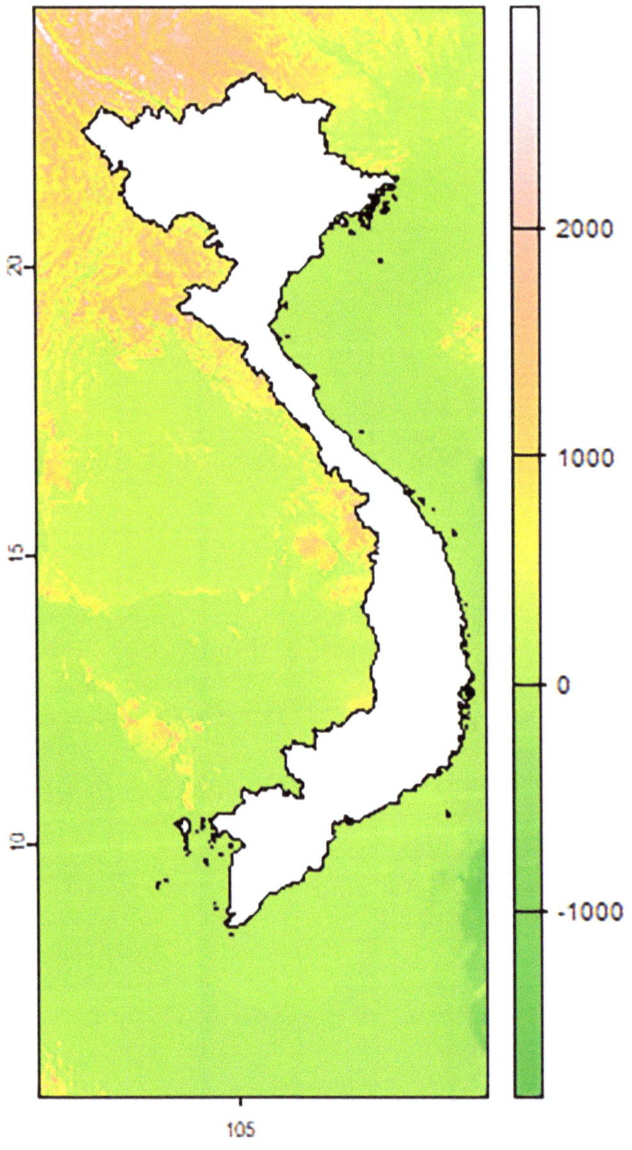

Fig. 6.3 A map with two layers: a raster layer (SpatRaster) with the SRTM elevation model and a vector layer (sfc_MULTIPOLYGON) with the Vietnam state border. Without a transformation, the vector data imported using the ***rgeoboundaries::gb_adm0()*** function are not fully appropriate for map preparation in this case

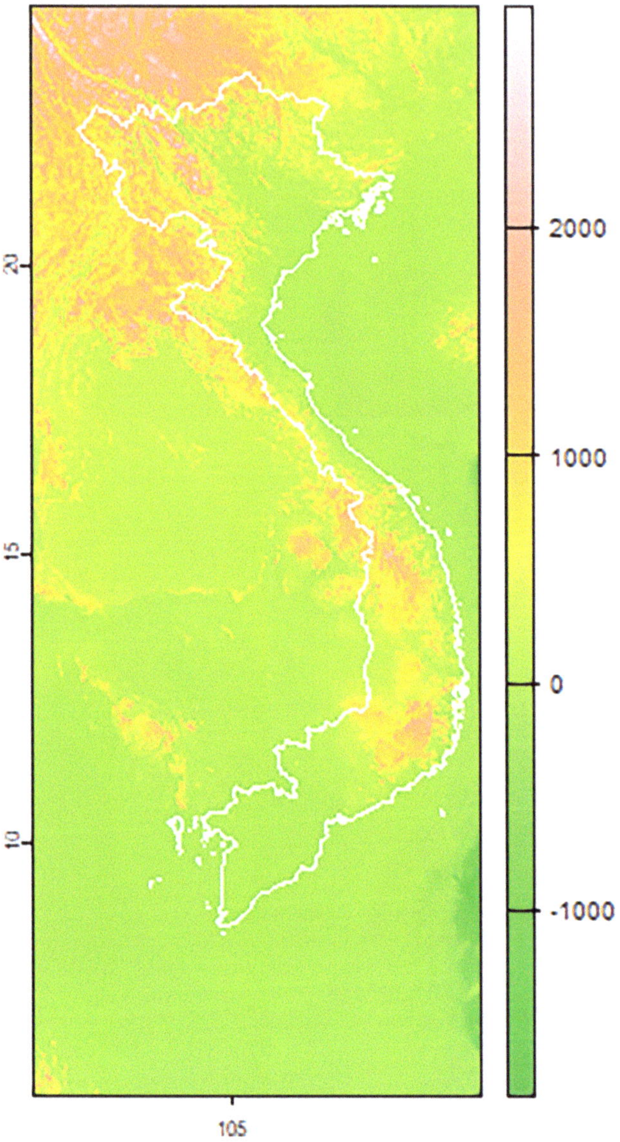

Fig. 6.4 After the transformation of the vector layer with the Vietnam state border, the elevation data for the Vietnam territory can be obtained. The data are transformed from MULTIPOLYGON to MULTILINESTRING form using the ***sf::st_cast()*** function

```
pacman::p_load("climenv", "rgeoboundaries",
"sf", "ggplot2", "dplyr",
 "raster", "terra", "tmap")
library(climenv) #access to spatial data sets
library(rgeoboundaries) #state borders as
polygons
library(sf) #spatial data tools
library(ggplot2)
library(dplyr)
#Importing spatial data for the Vietnam state
border.
vietnam <- gb_adm0("Vietnam")
#We can plot it using the base R plot()
function
#plot(vietnam$geometry)
```

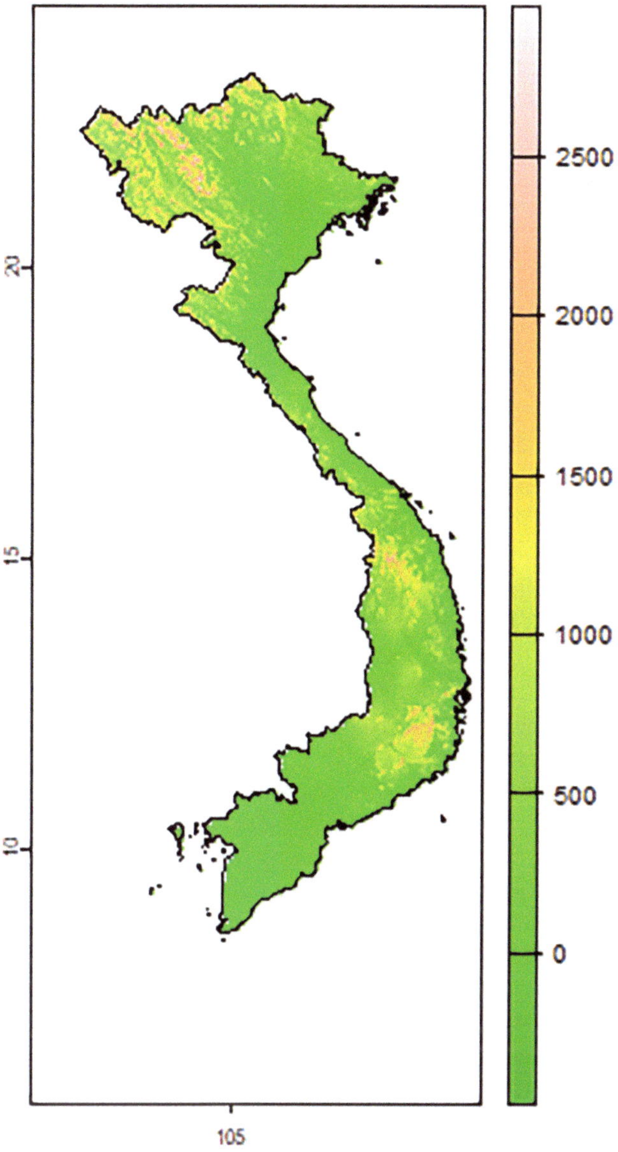

Fig. 6.5 The raster layer can be masked to the range of the vector layer. In this case, we reduce the quantity of data and focus on the information within the vector layer. This can be achieved using the ***terra::mask()*** function

```
#Import elevation data for the Vietnam
territory and save it in a temporary file.
temp_dir <- tempdir()
vt_elev <- elev(output_dir =temp_dir,
               location = vietnam)
#What is the object type?
vt_elev
#It is raster data in a SpatRaster class.
library(terra)
#The raster layer with the elevation data can
be plotted using the plot() function.
#We also add a polygon layer with the Vietnam
border.
#png("viet_raster1.png")
```

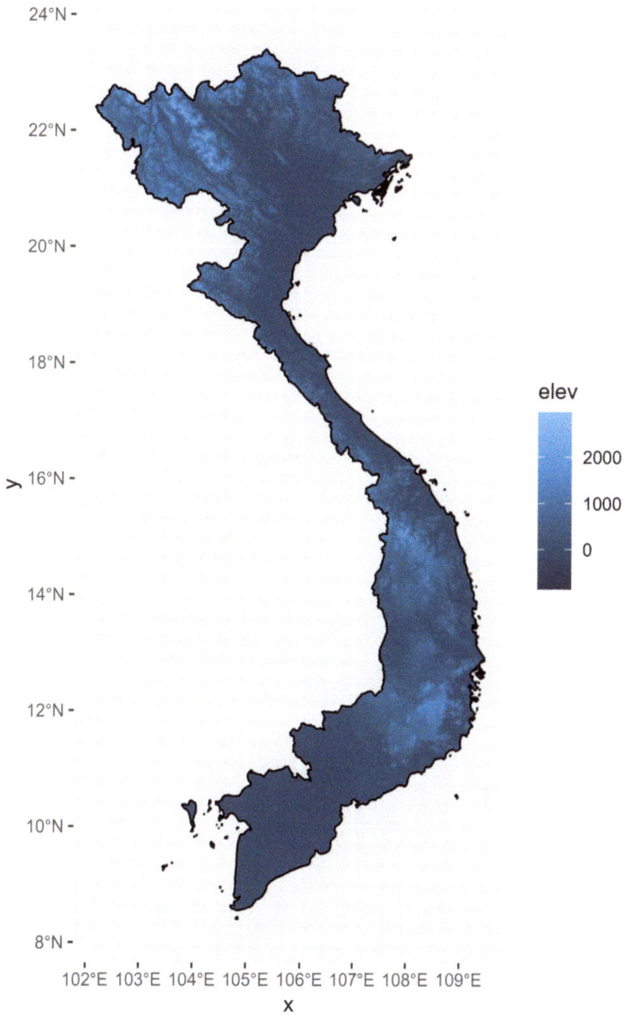

Fig. 6.6 The final map is generated using the **ggplot2** package with the default color scale

```
plot(vt_elev, col = terrain.colors(100)) #plot
elevation raster
plot(vietnam$geometry, col = "white", add =
TRUE) #add borderline
#dev.off()
#Why does it cover the digital elevation model?
#To show it differently, we must transform it
into a line.
vt_line <- st_cast(vietnam, "MULTILINESTRING")
#png("viet_raster2.png")
plot(vt_elev, col = terrain.colors(100)) #plot
elevation raster
plot(vt_line, col = "white", add = TRUE) #add
borderline
#dev.off()
#We can also mask the DEM with the mask()
function from the terra package.
vt_elev1 <- terra::mask(vt_elev, vietnam)
#And add the layer with a line indicating the
country border.
```

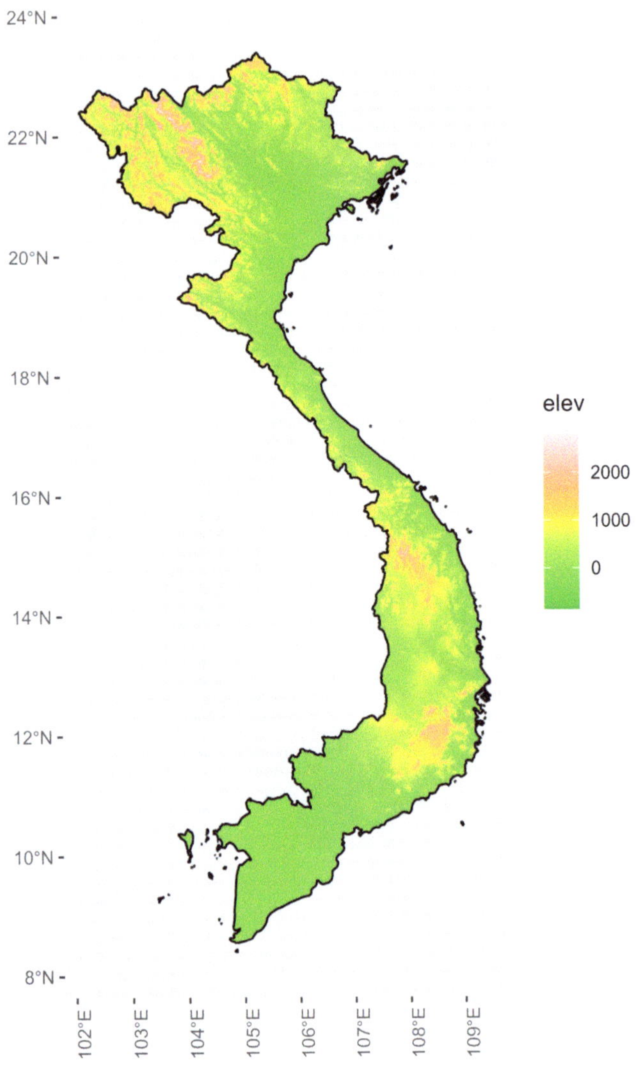

Fig. 6.7 The Vietnam elevation data are based on the SRTM model with a color scale dedicated to elevation visualization

```
#png("viet_raster3.png")
plot(vt_elev1, col = terrain.colors(100)) #plot
elevation raster
plot(vt_line$geometry, col = "black", add =
TRUE) #add borderline
#dev.off()
#We do the same but using the ggplot2 package.
vt_elev1_df <- as.data.frame(vt_elev1, xy =
TRUE)
#What is in this dataset?
head(vt_elev1_df)
#We change the name of the last column to
"elev".
colnames(vt_elev1_df) <- c("x", "y", "elev")
ggplot()+
  geom_raster(data=vt_elev1_df, aes(x=x, y=y,
  fill=elev))+
  geom_sf(data=vt_line)
#ggsave("vietnam_elev1.png", dpi = 600)
#The color scale can be modified.
ggplot()+
  geom_raster(data=vt_elev1_df, aes(x=x, y=y,
  fill=elev))+
  geom_sf(data=vt_line)+
  scale_fill_gradientn(colors = terrain.
  colors(10))+
  theme
(axis.title.x = element_blank(),
        axis.title.y = element_blank(),
        axis.text.x = element_text(angle = 90))
#ggsave("vietnam_elev2.png", dpi = 600)
```

6.2 R in Practice: Digital Landforms

With DEM, basic geomorphometric analyses can be conducted. This task can be accomplished with the ***terrain()*** function from the **terra** package, which requires a *SpatRaster* object as an input and is dedicated to computing **slope**, **aspect**, **TPI**, **TRI**, **TRIriley**, **TRIrmsd**, **roughness**, and **flowdir** (flow direction). The maps in Fig. 6.8 present general information about the complexity of the ground topography, and more in-depth analyses at the local scale should be supported by higher-resolution data and a set of different algorithms. In this case, the maps were constructed with the **tmap** package (Tennekes 2018). To show various possibilities of TPI computation, a smaller part of the Vietnam territory is selected and used in this example (Fig. 6.9). The TPI displays local variability with topography and can be used in relief classification with several possible classes characterizing local topography, such as flat surfaces, open slopes, upper slopes, ridges, mountain tops, shallow valleys, and headwaters (Mokarram et al. 2015). For this task, the ***TPI()*** function from the **MultiscaleDTM** package (Ilich et al. 2021) is used. Technically, during the calculation, the difference between the value of the focal cell and the mean values of the surrounding cells is determined at a given location (Ilich et al. 2023). The mean value is calculated within a predefined focal window (rectangular or circular). Another way to control how the TPI is calculated is changing the numbers of rows and columns used to define the size of the focal window.

```
#Terrain slope.
sl <- terra::terrain(vt_elev1)
aspect <- terra::terrain(vt_elev1, v="aspect")
#ground surface aspect
tpi <- terra::terrain(vt_elev1, v="TPI") #TPI -
Topographic Position Index
tri <- terra::terrain(vt_elev1, v="TRI") #TRI -
Terrain Ruggedness Index
```

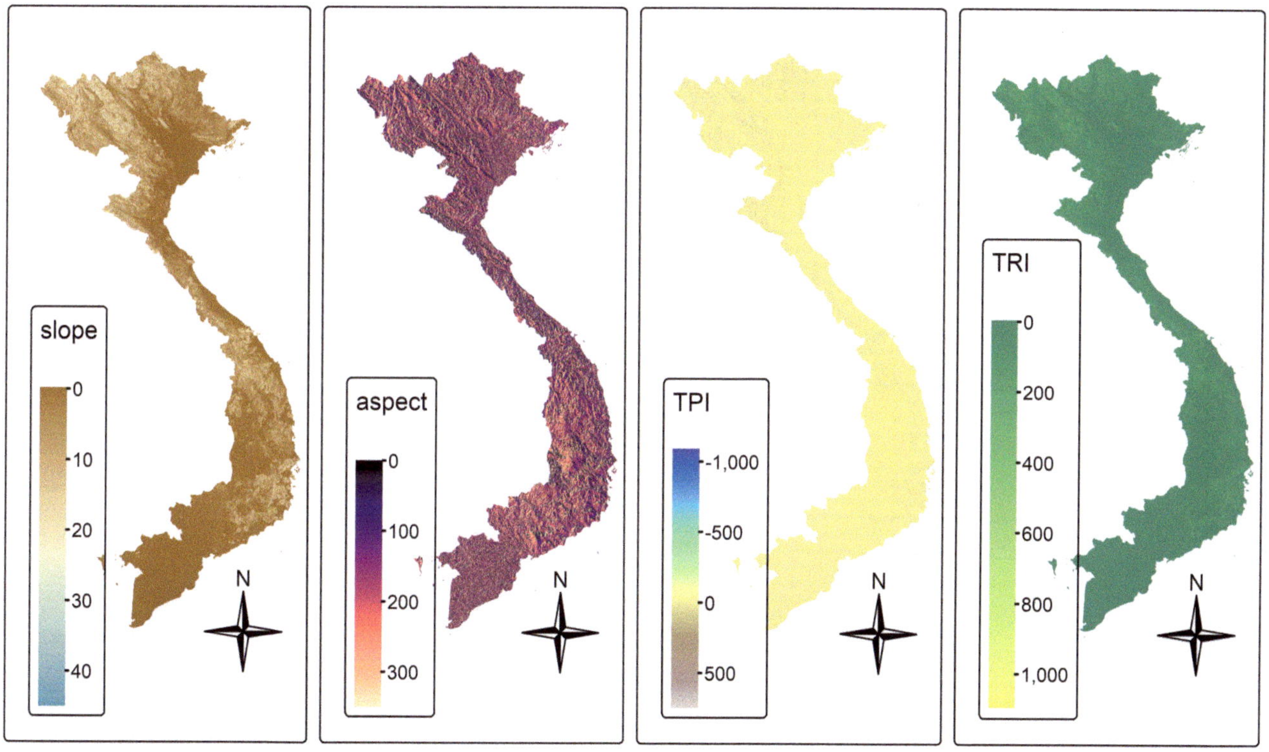

Fig. 6.8 Selected geomorphometric derivatives calculated for the Vietnam territory via the SRTM model. *TPI* topographic position index, and *TRI* terrain roughness index

```
library(tmap)
#Terrain slope map.
tmap_sl <- tm_shape(sl)+
  tm_raster(col.scale = tm_scale_
  continuous(values = "carto.earth"))+
  tm_compass(type = "4star", size = 3, position
  = c(0.6,0.25))+
  tm_layout(legend.position = tm_pos_in("left",
  "bottom"))
#Note the difference in color scales. You can
change the color scale values using the tm_
scale_continuous() function. Run the following
code to inspect different options for the
matplotlib color scale.
cols4all::c4a_palettes(type = "seq", series =
"matplotlib")
#For the maps below, we used different color
scales to show how to change the map style and
adapt it to a specific variable displayed on the
map.
#Aspect map.
tmap_aspect <- tm_shape(aspect)+
  tm_raster(col.scale = tm_scale_
  continuous(values = "matplotlib.magma"))+
  tm_compass(type = "4star", size = 3, position
  = c(0.6,0.25))+
  tm_layout(legend.position = tm_pos_in("left",
  "bottom"))
```

```
#TPI map.
tmap_tpi <- tm_shape(tpi)+
  tm_raster(col.scale = tm_scale_
  continuous(values = "matplotlib.terrain"))+
  tm_compass(type = "4star", size = 3, position
  = c(0.6,0.25))+
  tm_layout(legend.position = tm_pos_in("left",
  "bottom"))
#TRI map.
tmap_tri <- tm_shape(tri)+
  tm_raster(col.scale = tm_scale_
  continuous(values = "matplotlib.summer"))+
  tm_compass(type = "4star", size = 3, position
  = c(0.6,0.25))+
  tm_layout(legend.position = tm_pos_in("left",
  "bottom"))
#Arrange all maps.
tmap1 <- tmap_arrange(tmap_sl, tmap_aspect,
tmap_tpi, tmap_tri,
                nrow = 1, ncol = 4)
#Display them.
tmap1
#Save the figure with maps.
#tmap_save(tmap1, "tmap_maps.png", dpi = 600,
width = 20, units = "cm", asp = NA)

#Importing spatial data for Vietnam state
border and provinces.
```

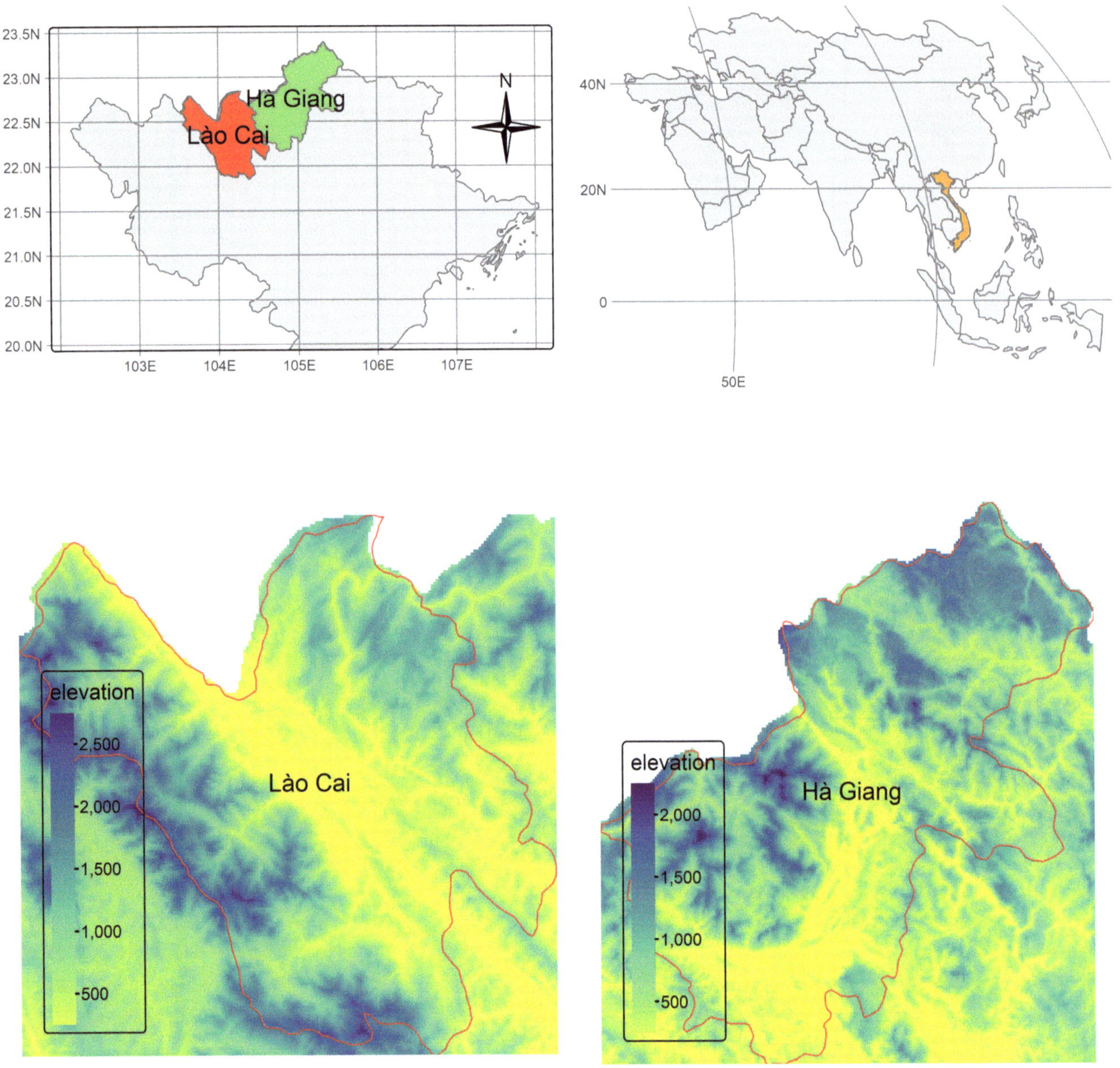

Fig. 6.9 Elevation data for two provinces in northern Vietnam: Lào Cai and Ha Giang

```
viet_regions <- gb_adm1("Vietnam")
#We select two provinces in northern Vietnam:
Ha Giang and Lao Cai.
#Although it is the "sfg" (simple feature
geometry) object, we can filter it using tools
dedicated to tabular data.
cai <- viet_regions[viet_regions$shapeISO ==
"VN-02",]
giang <- viet_regions[viet_regions$shapeISO ==
"VN-03",]
pacman::p_load("MultiscaleDTM")
#The elevation data is in the vt_elev object.
We use the borders of the two provinces to mask
the DEM.
library(MultiscaleDTM)
#Calculate TPI for different "rectangle" focal
windows.
tpi1 <- TPI(vt_elev, w=c(3,3))
tpi2 <- TPI(vt_elev, w=c(7,7))
tpi3 <- TPI(vt_elev, w=c(51,51))
#Crop the raster data to the vector extent -
Lào Cai.
cai_tpi1 <- terra::crop(tpi1, cai)
cai_tpi2 <- terra::crop(tpi2, cai)
cai_tpi3 <- terra::crop(tpi3, cai)
#Crop the raster data to the vector extent - Ha
Giang.
giang_tpi1 <- terra::crop(tpi1, giang)
```

```
giang_tpi2 <- terra::crop(tpi2, giang)
giang_tpi3 <- terra::crop(tpi3, giang)
#Visualization.
library(tmap)
data("World")#World map
#We select only the Asia continent.
asia_map <- tm_shape(World[World$continent ==
"Asia",], projection = 8857)+
  tm_polygons()+
  tm_shape(vietnam)+tm_polygons(fill="orange")+
  tm_layout(frame = NA)+
  tm_graticules()
bbox1 <- sf::st_bbox(c(xmin=102,xmax=108.1,ymin
=20,ymax=23.5))
map_vietnam1 <- tm_shape(vietnam, bbox = bbox1,
is.main = TRUE)+
  tm_polygons()+
  tm_graticules()+
  tm_shape(giang)+tm_polygons(fill =
  "green")+tm_text("shapeName")+

tm_shape(cai)+tm_polygons(fill = "red")+tm_
text("shapeName")+
  tm_compass(type = "4star", size = 3, position
  = c(0.8,0.9))
cai_elev <- terra::crop(vt_elev1, cai)
giang_elev <- terra::crop(vt_elev1, giang)
names(cai_elev) <- "elevation"
cai_elev_amp <- tm_shape(cai_elev)+
  tm_raster(col.scale = tm_scale_
  continuous(values = "viridis"))+
  tm_shape(cai)+tm_lines(col =
  "red")+tm_text("shapeName")+
  tm_layout(legend.position = tm_pos_in("left",
  "bottom"),
            legend.reverse = TRUE, frame = NA)
names(giang_elev) <- "elevation"
giang_elev_amp <- tm_shape(giang_elev)+
  tm_raster(col.scale = tm_scale_
  continuous(values = "viridis"))+
  tm_shape(giang)+tm_lines(col =
  "red")+tm_text("shapeName")+
  tm_layout(legend.position = tm_pos_in("left",
  "bottom"),
            legend.reverse = TRUE, frame = NA)
#Arrange the maps.
tmap2 <- tmap_arrange(map_vietnam1, asia_
map,cai_elev_amp, giang_elev_amp,
             ncol=2)
#Display the figure.
tmap2
#Save the figure.
#tmap_save(tmap2, "tmap_regions_elevation.png",
dpi = 600, width = 20, units = "cm", asp = NA)
```

Three focal window sizes were selected: 3 × 3, 7 × 7, and 51 × 51. The TPI 3 × 3 has the smallest focal window, and this approach revealed local differences in topography at short distances, not exceeding 300 m (Fig. 6.10). When the size of the focal window is too small, this method is not able to capture regional elevation changes between mountain ridges and valley bottoms. The TPI 51 × 51 model revealed differences in elevation up to 1500 m.

```
#Preparing maps with the TPI models.
# Lào Cai Province.
cai_map1 <- tm_shape(cai_tpi1)+
 tm_raster(col.scale = tm_scale_
continuous(values = "earth"))+
 tm_title("TPI 3x3")+
 tm_shape(cai)+tm_lines(col =
 "red")+tm_text("shapeName")+
 tm_layout(legend.position = tm_pos_in("left",
 "bottom"),
 title.position = tm_pos_in("right", "top"),
 legend.reverse = TRUE,
 legend.frame = NA)
cai_map2 <- tm_shape(cai_tpi2)+
 tm_raster(col.scale = tm_scale_
 continuous(values = "earth"))+
  tm_title_out("TPI 7x7")+
  tm_shape(cai)+tm_lines(col =
  "red")+tm_text("shapeName")+
   tm_layout(legend.position = tm_pos_
   in("left", "bottom"),
             title.position = tm_pos_
             in("right", "top"),
             legend.reverse = TRUE,
             legend.frame = NA)
cai_map3 <- tm_shape(cai_tpi3)+
  tm_raster(col.scale = tm_scale_
  continuous(values = "earth"))+
  tm_title_out("TPI 51x51")+
  tm_shape(cai)+tm_lines(col =
  "red")+tm_text("shapeName")+
   tm_layout(legend.position = tm_pos_
   in("left", "bottom"),
             title.position = tm_pos_
             in("right", "top"),
             legend.reverse = TRUE,
             legend.frame = NA)
#Ha Giang Province.
giang_map1 <- tm_shape(giang_tpi1)+
  tm_raster(col.scale = tm_scale_
  continuous(values = "earth"))+
  tm_shape(giang)+tm_lines(col =
  "red")+tm_text("shapeName")+
   tm_layout(legend.position = tm_pos_
   in("left", "bottom"),
```

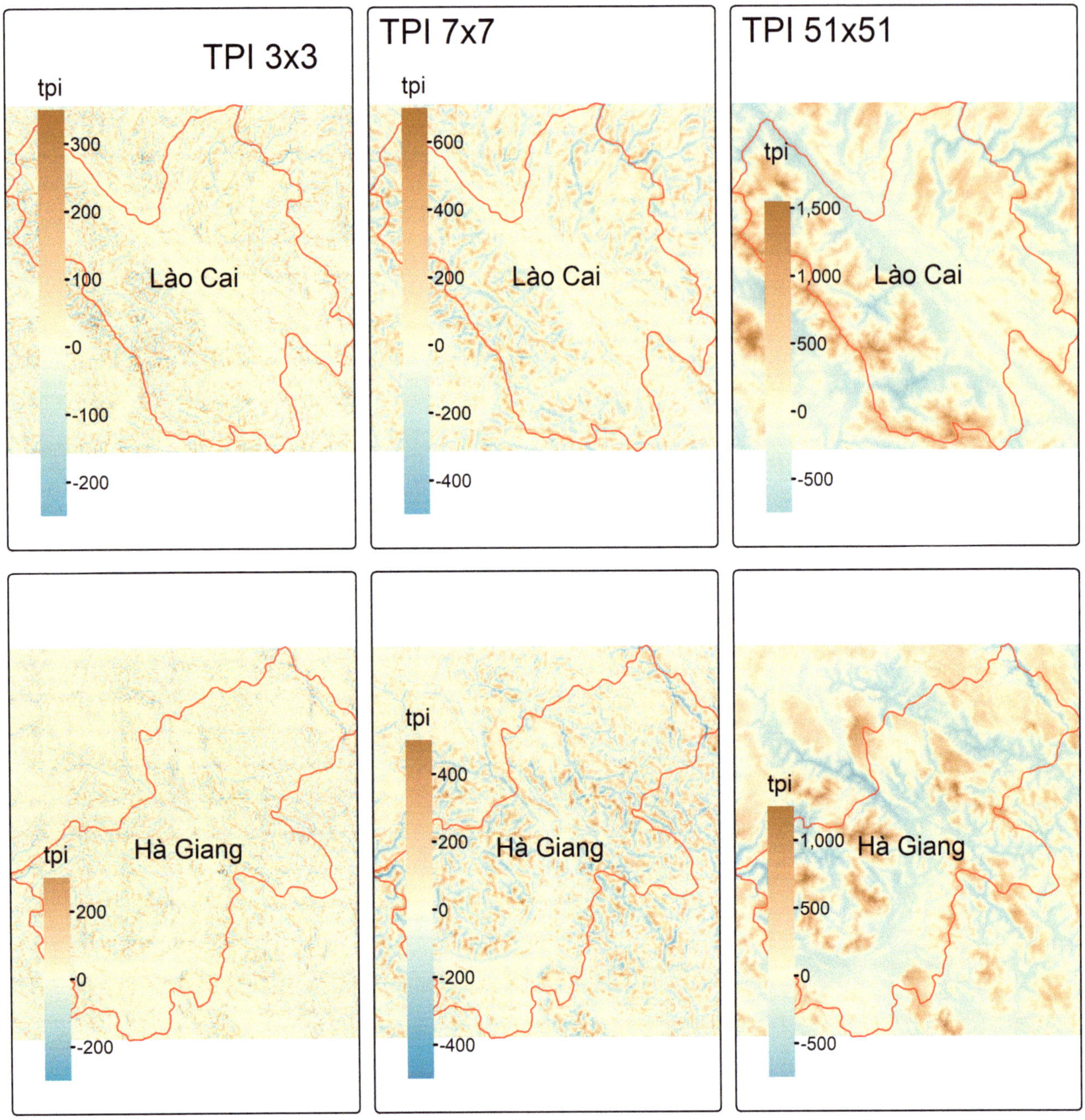

Fig. 6.10 The TPI parameter calculated for two provinces in Vietnam. Three sizes of focal windows were applied to show how the method works with different function arguments and how changes in topography and elevation are displayed when various optional parameters are adopted

```
        legend.reverse = TRUE,
        legend.frame = NA)
giang_map2 <- tm_shape(giang_tpi2)+
  tm_raster(col.scale = tm_scale_
  continuous(values = "earth"))+
  tm_shape(giang)+tm_lines(col =
  "red")+tm_text("shapeName")+
   tm_layout(legend.position = tm_pos_
   in("left", "bottom"),
```

```
        legend.reverse = TRUE,legend.frame
        = NA)
giang_map3 <- tm_shape(giang_tpi3)+
  tm_raster(col.scale = tm_scale_
continuous(values = "earth"))+
  tm_shape(giang)+tm_lines(col =
  "red")+tm_text("shapeName")+
   tm_layout(legend.position = tm_pos_
   in("left", "bottom"),
```

```
            legend.reverse = TRUE, legend.
            frame = NA)
#Arrange the maps.
tmap3 <- tmap_arrange(cai_map1, cai_map2,
cai_map3,
                      giang_map1, giang_map2,
                      giang_map3,
            ncol=3, nrow = 2)
#Display the maps.
tmap3
#Save the figure.
#tmap_save(tmap3, "tpi_different_versions.png",
dpi = 600)
```

References

Al-Sababhah N (2023) Topographic position index to landform classification and spatial planning, using GIS, for Wadi araba, South West Jordan. Environ Ecol Res 11:79–101. https://doi.org/10.13189/eer.2023.110106

Bloom AL (1969) The surface of the earth. Prentice-Hall, Inc., Englewood Cliffs, New Jersey

Chorley RJ, Kennedy BA (1971) Physical geography. A system approach. Prentice Hall International Inc., London

Dicko A (2024) rgeoboundaries: geoBoundaries Client

Diniega S, Bramson AM, Buratti B et al (2021) Modern Mars' geomorphological activity, driven by wind, frost, and gravity. Geomorphology 380:107627. https://doi.org/10.1016/j.geomorph.2021.107627

Dyderski MK, Pawlik Ł (2020) Spatial distribution of tree species in mountain national parks depends on geomorphology and climate. For Ecol Manag 474:118366. https://doi.org/10.1016/j.foreco.2020.118366

Evans IS (1972) General geomorphometry, derivatives of altitude, and descriptive statistics. In: Spatial analysis in geomorphology. Harper & Row, pp 17–90

Hijmans R (2023) terra: spatial data analysis. R Package Version 1(7–1):8

Huggett RJ (2007) Fundamentals of geomorphology. Routledge Taylor & Francis Group, London

Ilich AR, Misiuk B, Lecours V, Murawski SA (2021) Multiscale DTM

Ilich AR, Misiuk B, Lecours V, Murawski SA (2023) MultiscaleDTM: an open-source R package for multiscale geomorphometric analysis. Trans GIS 27:1164–1204. https://doi.org/10.1111/tgis.13067

Ironside KE, Mattson DJ, Arundel T et al (2018) Geomorphometry in landscape ecology: issues of scale, physiography, and application. Environ Ecol Res 6:397–412. https://doi.org/10.13189/eer.2018.060501

Lee SJ, Clarke JC (2005) An assessment of differences in algorithms for computing fundamental topographic parameters. Las Vegas, Nevada, USA, pp 1–21

Mokarram M, Roshan G, Negahban S (2015) Landform classification using topography position index (case study: salt dome of Korsia-Darab plain, Iran). Model Earth Syst Env 1:1. https://doi.org/10.1007/s40808-015-0055-9

Moore ID, Gessler PE, Nielsen GA, Peterson GA (1993) Soil attribute prediction using terrain analysis. Soil Sci Soc Am J 57:443–452. https://doi.org/10.2136/sssaj1993.03615995005700020026x

Olaya V (2008) Basic land-surface parameters. In: Geomorphometry: concepts, software, applications. Elsevier, Amsterdam, pp 141–169

Pawlik Ł, Harrison SP (2022) Modelling and prediciton of wind damage in forest ecosystems of the Sudety Mountains, SW Poland. Sci Total Environ 815:151972. https://doi.org/10.1016/j.scitotenv.2021.151972

Pawlik Ł, Samonil P, Malik I et al (2019) Geomorphic edge effects in response to abiotic and anthropogenic disturbances in forest ecosystems of the Gorce Mountains, Western Carpathians. Catena 177:134–148. https://doi.org/10.1016/j.catena.2019.02.013

Pawlik Ł, Buma B, Wistuba M et al (2023) Trees as bioindicators of hillslope degradation by debris flows and dangerous rockfalls along the Lefthand canyon, Colorado front range. Land Degrad Dev 34:1869–1884. https://doi.org/10.1002/ldr.4575

Pawlik Ł, Kasprzak M, Ignatiuk D et al (2024) Evaluation of the hillslope fine-scale morphology under forest cover with pit-mound topography—integration of geomorphometry, geophysical methods, and soil features. Geomorphology 460:109283. https://doi.org/10.1016/j.geomorph.2024.109283

Phillips JD (2009) Changes, perturbations, and responses in geomorphic systems. Prog Phys Geogr 33:17

Pike RJ, Evans IS, Hengl T (2008) Geomorphometry: a brief guide. In: Geomorphometry: concepts, software, applications. Elsevier, Amsterdam, pp 3–30

Pitty AF (1971) Introduction to geomorphology. Methuen & Co Ltd, London

Riley SJ, DeGloria SD, Elliot R (1999) A terrain ruggedness index that quantifies topographic heterogeneity. Int J Sci 5:23–27

Rodriguez E, Morris CS, Belz JE et al (2005) An assessment of the SRTM topographic products. Jet Propulsion Laboratory, Pasadena

Temme AJ, Schoorl JM, van der Meij WM (2022) New geomorphometric variables for non-continuous hillslopes—assessing the value for digital soil mapping. Geoderma 418:115848. https://doi.org/10.1016/j.geoderma.2022.115848

Tennekes M (2018) tmap: thematic maps in R. J Stat Softw 84:1–39. https://doi.org/10.18637/jss.v084.i06

Tsakalos JL, Smith MR, Luebert F, Mucina L (2023) climenv: download, extract and visualise climatic and elevation data. J Veg Sci 34:1–7. https://doi.org/10.1111/jvs.13215

Wilson MFJ, O'Connell B, Brown C et al (2007) Multiscale terrain analysis of multibeam bathymetry data for habitat mapping on the continental slope. Mar Geod 30:3–35. https://doi.org/10.1080/01490410701295962

Xiong L, Li S, Tang G, Strobl J (2022) Geomorphometry and terrain analysis: data, methods, platforms and applications. Earth Sci Rev 233:104191. https://doi.org/10.1016/j.earscirev.2022.104191

Dating the Past with the Radiocarbon Method

7

Słowiński National Park, Poland, 2020, photo: T. Pawlik

'To the past', said Winston.
'The past is more important', agreed O'Brien gravely.
George Orwell, Nineteen Eighty-Four (1949)

In this chapter, we are interested in dating past events by analyzing fossil materials. Among the many techniques used in the Earth sciences, one of the most popular is the **radiocarbon dating** method (Bowman 1990; Walker 2005; Hajdas 2008; Wood 2015). The method was proposed by Willard Libby (1908–1980), a professor of chemistry at the University of Chicago, in 1946. For his discovery, in 1960, he obtained a Nobel Prize (Libby 1960; Olsson 2009). Owing to this discovery, another revolution in the Earth sciences commenced—the radiocarbon revolution.

Naturally occurring carbon has three isotopes, two stable (^{12}C and ^{13}C) and one unstable (^{14}C), that decays spontaneously

Supplementary Information The online version contains supplementary material available at https://doi.org/10.1007/978-3-031-89673-6_7.

© The Author(s), under exclusive license to Springer Nature Switzerland AG 2025

Ł. Pawlik, *R Applications in Earth Sciences*, Springer Textbooks in Earth Sciences, Geography and Environment,
https://doi.org/10.1007/978-3-031-89673-6_7

after some time (Walker 2005). The two stable and most common isotopes, ^{12}C and ^{13}C, represent 98.9% and 1.1%, respectively, of all naturally occurring carbon (Walker 2005; Hajdas 2008). The quantity of ^{14}C isotopes in the atmosphere is very low. If expressed as the $^{14}C/^{12}C$ ratio, it is only 10^{-12} (Hajdas et al. 2021). The radiocarbon method relies on the unstable carbon isotope ^{14}C, which decays after a so-called half-life of 5700 ± 30 years (Hajdas et al. 2021). Willard Libby discovered that by counting the decay rate of ^{14}C atoms, the real age of a carbon-bearing material could be assessed. Because every living organism accumulates a fraction of carbon through food consumption or photosynthesis, some tiny part of this carbon consists of ^{14}C. Before any material consisting of radiocarbon is dated, the ^{14}C isotope travels through the upper atmosphere (the zone of its production) and the bodies of living organisms and is captured in a variety of fossils that are sampled and transported to laboratories. After the death of a living organism, the radiocarbon clock starts. There is no longer a way to maintain the balance between stable and unstable carbon isotopes. ^{14}C decays slowly following its half-life decay rate. Willard Libby described this phenomenon in his Nobel lecture in the following words (Libby 1960):

> *(...) organic matter, while it is alive, is in equilibrium with the cosmic radiation, (...), all the radiocarbon atoms which disintegrate in our bodies are replaced by the carbon-14 contained in the food we eat, so that while we are alive we are part of a great pool which contains the cosmic-ray produced radiocarbon.*
>
> *(...) At the time of death, however, the assimilation process stops abruptly. There is no longer any process by which carbon-14 from the atmosphere can enter our bodies. (...)*

This information leads us to define the data type we need to determine the age of carbon-bearing materials:

1. The rate of ^{14}C at a specific time in the past (the time the organism was still alive and could assimilate carbon isotopes, including ^{14}C).
2. The half-life of ^{14}C. The half-life rate was tested against historical material of known age (Fig. 7.1).

The ^{14}C isotope is produced by cosmic ray bombardment in the atmosphere, the production of new neutrons, and their reaction with nitrogen (^{14}N). With an extra neutron, ^{14}N (7 protons +7 neutrons) is transformed into radiocarbon ^{14}C (with 6 protons and 8 neutrons). This production was calculated for 2.02 atoms cm^{-2} s^{-1} of ^{14}C (Libby 1960; Hajdas 2008). After oxidation to CO and CO_2, ^{14}C is mixed in the atmosphere and assimilated by plants through photosynthesis, and through gas exchange, it enters the ocean (Hajdas 2008; Höflmayer 2014). There are two methods used in dif-

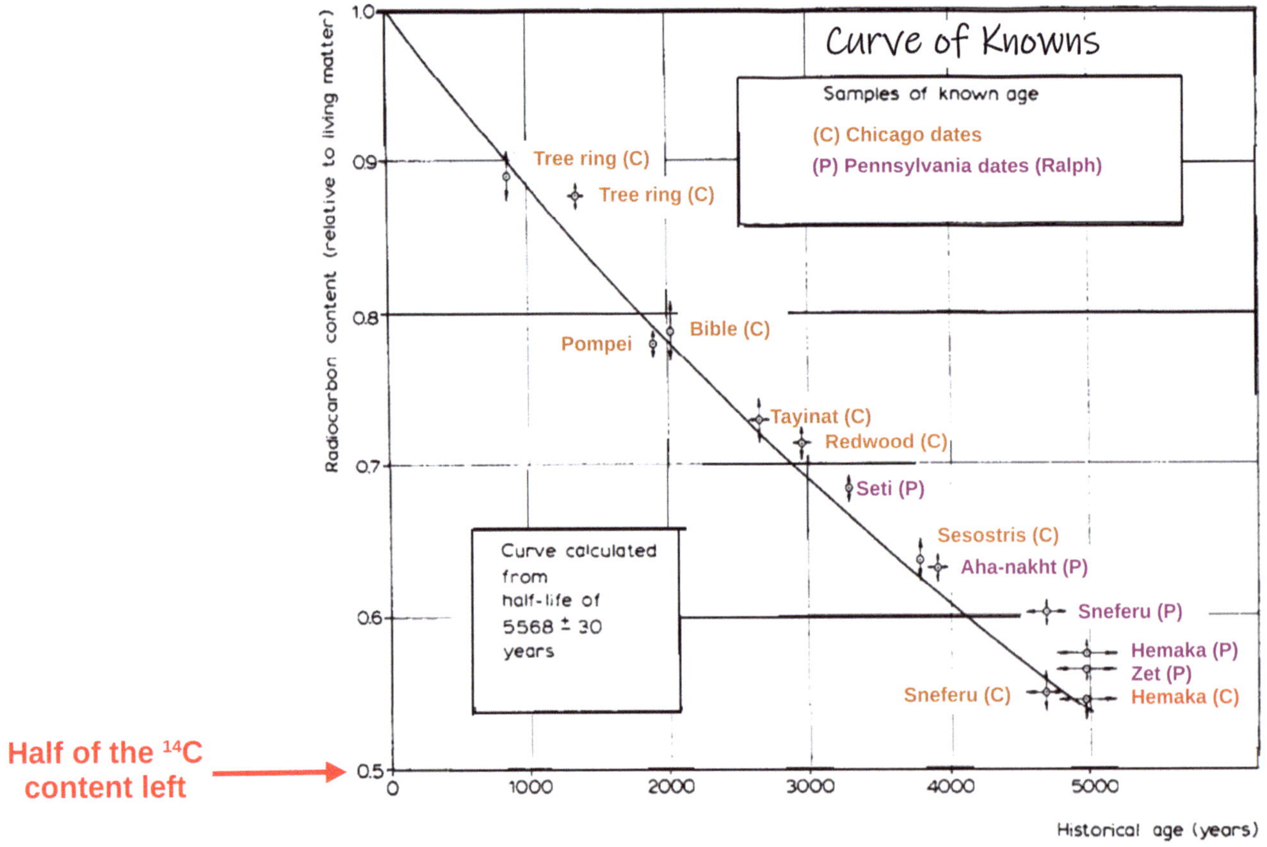

Fig. 7.1 Curve of Knowns. The results obtained for samples of known age compared with the carbon-14 decay curve, assuming that the present level of ^{14}C in living matter is equal to unity, with a half-life of 5568 ± 30 years (Libby 1960). The "Pompei" sample was measured by E.A. Olson and W.S. Broecker of the Lamont Geological Observatory

ferent laboratories worldwide to determine ^{14}C levels: 1. beta counting (β-counting) and 2. Accelerator Mass Spectrometry (AMS) (Walker 2005). The first method relies on the fact that during decay, β electrons are released from ^{14}C atoms over some time (Walker 2005; Hajdas et al. 2021). The activity of β electrons can be measured and interpreted as an "age" by using gas proportional counters (GPCs) or liquid scintillation counters (LSCs). The method is based on counting (measuring) ^{14}C decay products, whereas the second method, AMS, is based on the direct counting of ^{14}C atoms that are detected on the basis of their atomic weights (Walker 2005; Povinec et al. 2009). Because the proportion of radiocarbon in the atmosphere is very small and its half-life promotes spontaneous decay, this method has a limit of up to ~55,000 years.

Information about radiocarbon age is very difficult to obtain and involves a multistage process of careful sample collection, preparation, and dating. Samples should be collected while fully considering the environmental context and material, which are important for accurate dating. This approach may include preserving horizons in soil profiles, the position of charcoals, or species of dated plant macrodetritus. Additionally, samples must not be contaminated before final dating. Although they are preprocessed in the laboratory and physically and chemically treated to eliminate contamination, there is always some probability that a sample contains fragments of younger material. Another issue that makes radiocarbon dating data unique and difficult to obtain in large quantities is the commercial price of dating services, especially AMS dating. In addition, a dating sample is burned during the procedure, and for this reason, the process cannot be repeated. In the case of some rare or historical materials, the chances of obtaining full information about the age of the sample can be relatively low.

7.1 Description of the Data and Project Outline

The example project is based on organic samples collected from two localities within the landslide body in the Wyspowy Beskidy Mountains in Poland (Pawlik et al. 2022). This landslide is complex and of unknown age. Currently, its area is covered by forest and features a network of first-order streams incising the landslide body to a depth of several meters in some localities (Fig. 4.2, Section 4). A fossilized piece of tree stem was found within the bank of one of these streams (Fig. 7.2). Two discs were cut off using a hand saw and transported to the laboratory for further analyses. For radiocarbon dating, two samples of wood were taken from the outer parts of the discs. The annual tree rings were still clearly visible, so it was possible to determine where the central part of the tree was located and where older tree rings were to be expected. The main concern was not to date the age of the tree but rather to determine the minimum age of the landslide event. Because the tree was buried in the colluvium deposits, it was hypothesized that the radiocarbon age of the trees could be interpreted as the minimum age of the landslide. For this reason, the wood from the oldest part of the tree was selected for radiocarbon dating (Figure 7.2a). The tree age was determined through dendrochronology (by measuring the number of tree rings). Two samples of fossil wood were dated to obtain specific results. If one sample from the same material was significantly younger or older than the other sample was, it would indicate contamination of the sample. The dated tree was a Scots pine (*Pinus sylvestris*) that was older than 100 years.

The second locality where the organic material was collected and dated was the same location from which samples for the geochemical analyses described in Sect. 4 were taken. Several radiocarbon dates from this peat core were used for the establishment of an accumulation chronology and rate. In total, 21 parts of the plants (needs, seeds, wood, and twigs) were dated. Only 13 dates were selected for the analyses presented in the next subsection because some dates did not follow the expected relationship with depth, i.e., older samples were positioned below younger samples along the profile.

7.2 R in Practice: Radiocarbon Data and Geochronology

Two assumptions are made when determining the conventional radiocarbon age, and for these reasons, calibration is required (Scott and Raimer 2009):

1. Constant rate of ^{14}C formation in the upper atmosphere.
2. Equilibrium exists among ^{14}C activity in the atmosphere, biosphere, and oceans.

The results obtained from the laboratory are conventional radiocarbon ages reported as Before Present (BP), where 0 BP = 1950 AD (Hajdas 2008). A constant atmospheric ^{14}C content is assumed, an assumption that is far from realistic. The differences in ^{14}C production are due to solar sunspot activity and geomagnetic modulation (Scott and Raimer 2009). This natural long-term fluctuation is also impacted by two anthropogenic effects: the combustion of fossil fuels and nuclear weapon testing (Scott and Raimer 2009). To obtain precise information on the real age of a carbon-bearing material, calibration is needed. What is calibration? This method is a sort of compensation for the error introduced by the conventional half-life and the temporal variability of the atmospheric ^{14}C content. Calendar or calibrated ages are expressed as "cal BP" (Reimer et al. 2020). Technically, calibration is a mathematical procedure that transforms radiocarbon ages into calibrated ages using a calibration curve constructed via other methods, such as dendrochronological

Fig. 7.2 Images taken during a field survey and sampling campaign. (**a**) Slice of buried pine stem that is almost 14,000 years old, (**b**) position of the stem before its excavation (the bottom of a river valley side), (**c**) cross-section of the river valley with the position of the pine stem before excavation (white arrow), and (**d**) the pine stem after its excavation. All images: Ł. Pawlik

or Uranium series methods for corals and speleothems (https://intcal.org/).

In R, there are tools available to conduct various tasks. First, the ***calibrate()*** function from the **rcarbon** package is applied (Crema and Bevan 2021). This step is necessary to transition from the radiocarbon age to the calibrated dates. However, in the case of the calibrated age, it is impossible to obtain an exact date, and we can obtain only the probability that the real age lies within the range of certain dates. However, if we consider a 68.2% probability acceptable, there is still a large chance that the real age lies outside the range of these dates (Table 7.1). The task is performed using an INTCAL20 (https://www.cambridge.org/core/journals/radiocarbon/article/intcal20-northern-hemisphereradiocarbon-age-calibration-curve-055-cal-kbp/83257B63DC3AF9CFA6243F59D7503EFF) calibration curve prepared by the International Calibration Group (Reimer et al. 2020). Another function that is dedicated to calibration is the ***BchronCalibrate()*** function from the **Bchron** package (Haslett and Parnell 2008).

Table 7.1 Calibration of radiocarbon ages displayed in Fig. 7.3. There is a 95% probability that the real age is between 2501 and 2753 years calBP

14C age	Median	68.2% probability	95.4% probability
2560 ±30 BP	2715	2746 to 2705	2753 to 2697
		2628 to 2620	2637 to 2615
		2555 to 2545	2592 to 2514
			2505 to 2501

```
pacman::p_load("rcarbon", "gt", "Bchron",
"tidyr")


#Another way to install the rcarbon package:
#devtools::install_github("ahb108/rcarbon")
```

```
library(rcarbon)


#The dataset with radiocarbon dates.


snieznica_C14 <- calibrate(x = c(330, 2560,
2730, 2885, 2940, 3005, 3225,
 3855, 3385, 3400, 3510, 3475, 3505, 3480,
 3315, 3570, 3825, 3985, 3895, 3760, 3950),
 errors = c(30, 30, 30, 35, 35, 30, 30, 30, 35,
 35, 35, 35, 35, 35, 35, 35, 35,
 35, 90, 90),
 calCurves = 'intcal20',
 ids = c('S096', 'S155', 'S168', 'S183',
'S200',
 'S238', 'S275', 'S276', 'S283', 'S303',
 'S323', 'S340', 'S341', 'S349', 'S350',
 'S360', 'S386', 'S393', 'S396', 'S405',
 'S422'))

#Present just one date.


one_date <- calibrate(x = c(2560),
 errors = c(30),
 calCurves = 'intcal20',
 ids = c('S155'))

#Plot the calibrated age probability.
plot(one_date)

#Save the plot.
#png("one_date_calibration_plot.png")
plot(one_date, main = "Mt Śnieżnica - 2560 ±30
BP")
#dev.off()

#Summary of the calibration procedure.
onedate_sum <- summary(one_date)

#And in a table form.
table_one_date <- data.frame(age14C = c("2560
±30 BP", "", "", ""),
                             median =
c(onedate_sum$MedianBP,"","",""),
                             onesigma =
c(onedate_sum$OneSigma_BP_1,

onedate_sum$OneSigma_BP_2,

onedate_sum$OneSigma_BP_3,
                                                  ""),
                             twosigma =
c(onedate_sum$TwoSigma_BP_1,

onedate_sum$TwoSigma_BP_2,

onedate_sum$TwoSigma_BP_3,

onedate_sum$TwoSigma_BP_4))

#With modified column names, the table can be
more informative.
colnames(table_one_date) <- c("14C age",
"median", "68.2% probability", "95.4%
probability")

#Show the table.
table_one_date

#Save the table using the gt library.


library(gt)


#gtsave(gt(table_one_date), "one_date_table.
png")


#Cumulative probability.
sn_spd <- spd(snieznica_C14, timeRange =
c(4000, 100))

#png("summed_probability.png")
plot(sn_spd, main = "Summed probability
distributions of radiocarbon dates")
#dev.off()
```

Figure 7.3 shows the calibration of radiocarbon ages. The left-hand Y axis displays radiocarbon determination, which is the amount of radiocarbon in a sample. To calibrate the sample, this amount needs to be compared with the amount in a sample for which the age is known (for example, tree rings) (https://c14.arch.ox.ac.uk/explanation.php). The yellow line represents the measurement of a known-age material, and the solid gray distribution represents the likelihood of different possible ages.

Importantly, the radiocarbon age of 2560 ± 30 BP is in the range of the so-called Hallstatt plateau, when reduced atmospheric ^{14}C activity was recorded c. 2520–2420 BP (c. 750–420 BC). This is problematic for precise dating because similar radiocarbon ages may yield calibrated ages spanning up to 500 calendar years (Hajdas 2008; Rose et al. 2022).

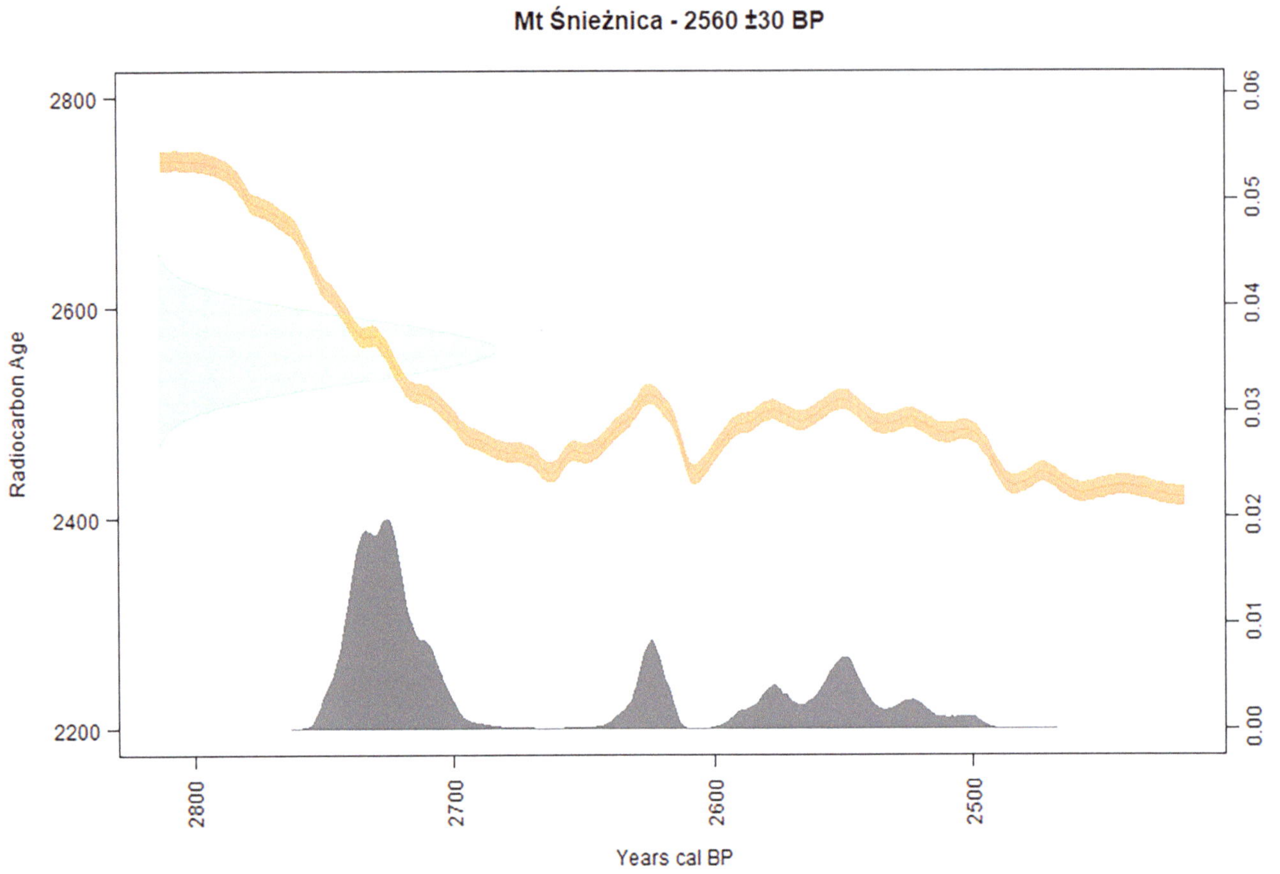

Fig. 7.3 Calibration of the radiocarbon age. The plot contains three types of information: 1. radiocarbon determination (left-hand axis), 2. the likelihood of different ages (x-axis) and 3. the yellow line, which is a calibration curve

Another way to look at calibrated radiocarbon dates is to use summed probability distributions. Individual ^{14}C calibrations are combined by adding the height of each calibrated distribution at every calendar date and normalizing the height of the cumulative probability density function (Contreras and Meadows 2014). Figure 7.4 shows all the data used for the task as a summed probability distribution. More comparatively older dates resulted in denser coverage of this period by the information obtained from radiocarbon dating.

After the depth and age of the analyzed samples are determined, the next two tasks can be accomplished by constructing an age–depth model and an accumulation rate diagram (Fig. 7.5). In these steps, we use the ***Bchronology()*** function from the **Bchron** package (Haslett and Parnell 2008). First, a nonparametric chronology model is constructed on the basis of the age and position of the samples (Fig. 7.6). Second, the chronological result is used to construct the accumulation rate plot. The plot provides information about the maximum depth of sampling, the age range of all samples, and the density of samples with known ages. There is a striking difference that occurs at c. 2500 calBP at a depth of approximately 150 cm. This is close to the boundary between peat sedimentation, which dominated below the 2-m depth, and lacustrine sedimentation, which features a younger layer. In this layer, no sample was found to be appropriate for dating purposes. These data impacted the sedimentation rate curve. According to the results, there were two peaks for which the sedimentation rate reached 0.4 cm year^{-1}. There was almost no sedimentation in the period after 2500 calBP.

```
library(Bchron)


#Calibration using the BchronCalibrate()
function from the Bchron package.


snieznica_C14a <- BchronCalibrate(ages = c(330,
2560, 2730, 2885, 2940, 3005, 3225, 3385, 3400,
3510, 3570, 3825, 3985),
 ageSds = c(30, 30, 30, 35, 35, 30, 30, 35, 35,
35, 35, 35, 35),
 calCurves = c('intcal20', 'intcal20',
'intcal20','intcal20', 'intcal20',
```

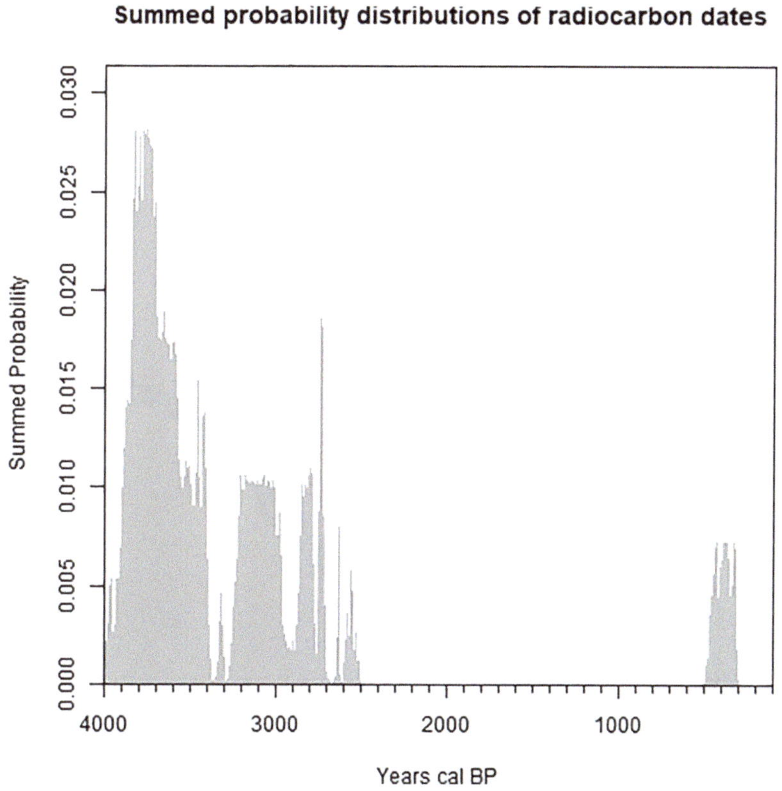

Fig. 7.4 Summed probability distribution of radiocarbon dates from the Mt. Śnieżnica study site

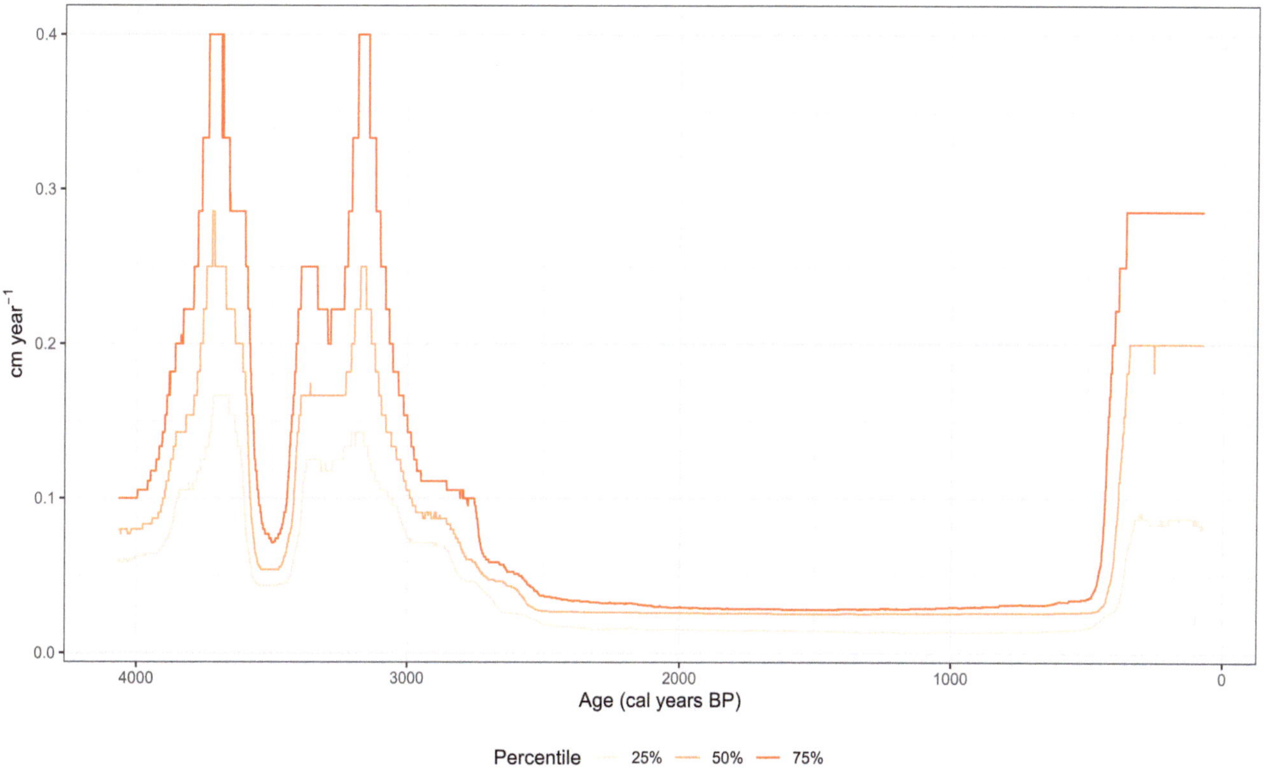

Fig. 7.5 Accumulation rate of deposits calculated against the calibrated radiocarbon age

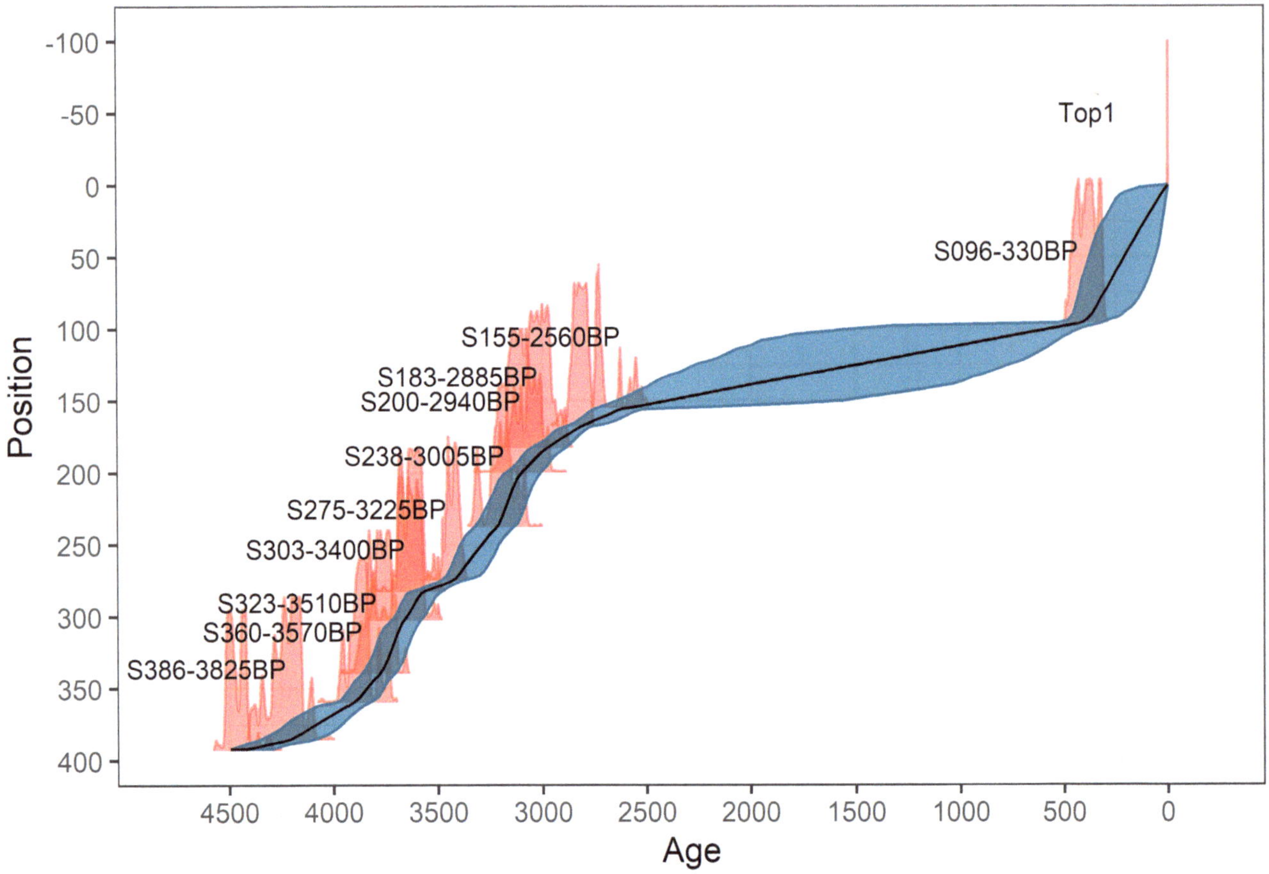

Fig. 7.6 Chronology model constructed on the basis of the radiocarbon ages of organic samples from the Mt. Śnieżnica study site, S Poland

```
 'intcal20', 'intcal20', 'intcal20',
'intcal20', 'intcal20',
'intcal20', 'intcal20', 'intcal20'),
 ids = c('S096-330BP', 'S155-2560BP', 'S168-
2730BP', 'S183-2885BP',
          'S200-2940BP','S238-3005BP', 'S275-
          3225BP', 'S283-3385BP',
           'S303-3400BP','S323-3510BP', 'S360-
           3570BP', 'S386-3825BP',
           'S393-3985BP'),
positions = c(96, 155, 168, 183, 200, 238, 275,
283, 303, 323, 360, 386, 393),
             allowOutside = TRUE)

#We can get a summary of the calibrated dates.
#summary(snieznica_C14)

#We can print a list of lists where each
element corresponds to a single age.
#print(snieznica_C14a)

library(dplyr)
library(rappliES)
data("radiocarbon")


#Fitting a non-parametric chronology model to
age/position data.
#We use the age-depth model proposed by Haslett
and Parnell (2008) and implemented in the
Bchron package.

snC14_chronology <- Bchronology(ids =
radiocarbon[,1],
         ages = radiocarbon[,2],
         ageSds = radiocarbon[,3],
         positions = radiocarbon[,4],
         positionThicknesses = radiocarbon[,5],
         calCurves = radiocarbon[,6],
         predictPositions = seq(0, 393, by=2))

#Plot and save the model.
#png("age_depth_model.png", units = 'cm', width
= 15, height = 10, res = 600)
plot(snC14_chronology, dateLabelSize = 3,
dateCol = rgb(240/255,
                                         59/255,
```

```
                                        32/255,
                                          0.5))
#dev.off()

#The object can be summerized using the
accumulation rate option.
acc_rate <- summary(snC14_chronology, type =
'acc_rate',
                    probs = c(0.25, 0.5, 0.75))

#The tidyr package helps to transform the table
from wide to long.
library(tidyr)

#Table transformation.
acc_rate_2 = acc_rate %>%
  pivot_longer(names_to = 'Percentile',
               values_to = 'value',
               -age_grid)

#Plotting.


library(ggplot2)


ggplot(acc_rate_2, aes(x = age_grid,
                       y = value,
                       colour = Percentile)) +
  scale_colour_brewer(palette = 7) +
  geom_line() + theme_bw() +
  scale_x_reverse() +
  labs(y = expression(cm ~ year^-1),
       x = 'Age (cal years BP)')+
  theme
(legend.position = 'bottom')

#Saving the plot.
#ggsave("accu_rate.png", dpi = 600, height = 6,
width = 10)
```

References

Bowman S (1990) Radiocarbon dating. University of California Press, Berkeley and Los Angeles

Contreras DA, Meadows J (2014) Summed radiocarbon calibrations as a population proxy: a critical evaluation using a realistic simulation approach. J Archaeol Sci 52:591–608. https://doi.org/10.1016/j.jas.2014.05.030

Crema ER, Bevan A (2021) Inference from large sets of radiocarbon dates: software and methods. Radiocarbon 63:23–39. https://doi.org/10.1017/RDC.2020.95

Hajdas I (2008) Radiocarbon dating and its applications in quaternary studies. Quater Sci J 57:2–24

Hajdas I, Ascough P, Garnett MH et al (2021) Radiocarbon dating. Nat Rev 1:1. https://doi.org/10.1038/s43586-021-0058-7

Haslett J, Parnell AC (2008) A simple monotone process with application to radiocarbon-dated depth chronologies. J R Stat Soc 57:399–418

Höflmayer F (2014) Radiocarbon dating and Egyptian chronology—from the "curve of knowns" to Bayesian modeling. In: The Oxford handbook of topics in archaeology. Oxford Handbooks Editorial Board

Libby WF (1960) Radiocarbon dating

Olsson IU (2009) Radiocarbon dating history: early days, questions, and problems met. Radiocarbon 51:1–43

Pawlik Ł, Okupny D, Kroh P et al (2022) Changing natural conditions and their impact on Mt Śnieżnica landscape, outer Western Carpathians—reconstruction of the Holocene environment based on geochemical and radiocarbon indices. Sci Total Environ 850:158066. https://doi.org/10.1016/j.scitotenv.2022.158066

Povinec PP, Litherland AE, von Reden KF (2009) Developments in radiocarbon technologies: from the Libby counter to compound-specific AMS analyses. Radiocarbon 51:45–78

Reimer PJ, Austin WEN, Bard E et al (2020) The INTCAL20 northern hemisphere radiocarbon age calibration curve (0-55 CAL kBP). Radiocarbon 62:725–757. https://doi.org/10.1017/RDC.2020.41

Rose HA, Müller-Scheeβel N, Meadows J, Hamann C (2022) Radiocarbon dating and Hallstatt chronology: a Bayesian chronological model for the burial sequence at Dietfurt an der Altmühl "Tennisplatz", Bavaria, Germany. Archaeol Anthropol Sci 14:72. https://doi.org/10.1007/s12520-022-01542-1

Scott EM, Raimer PJ (2009) Calibration introduction. Radiocarbon 51:283–285

Walker M (2005) Quaternary dating methods. John Wiley & Sons, New York

Wood R (2015) From revolution to convention: the past, present and future of radiocarbon dating. J Archaeol Sci 56:61–72. https://doi.org/10.1016/j.jas.2015.02.019

8 Global Tectonics and Earthquake Dynamics: From Data to Visualization

Serra da Canastra, Brazil, 2024, photo: Ł. Pawlik

Earthquakes, or tlalollin in the original Nahuatl language, are represented by two associated glyphs: movement or ollin and Earth or tlalli.
(Suárez and García-Acosta 2021)

Supplementary Information The online version contains supplementary material available at https://doi.org/10.1007/978-3-031-89673-6_8.

From the local scale of a single study site, we now transition to a global scale and continue our journey through various Earth science data. We live on unstable ground, and daily information that reaches us through the media confirms this statement. Earthquakes and volcanic eruptions

© The Author(s), under exclusive license to Springer Nature Switzerland AG 2025

Ł. Pawlik, *R Applications in Earth Sciences*, Springer Textbooks in Earth Sciences, Geography and Environment,
https://doi.org/10.1007/978-3-031-89673-6_8

cause damage to man-made infrastructure and even human deaths due to processes such as building collapse and the dissemination of volcanic ash. Earthquakes in southern Turkey and northern Syria in 2023 resulted in >52,000 deaths (Zilio and Ampuero 2023). In 2024, a wave of volcanic activity was recorded in Iceland. These natural phenomena are not surprising in these regions because both are tectonically active areas (Gudmundsson et al. 2008). Continents and oceanic bottoms are subject to continuous changes that are most significant along their boundaries where some processes are driven by plate tectonics (Niu 2014). Current measurements confirm that tectonic plates change their positions and that at the geological time scale, plates drift from the Southern Hemisphere to the Northern Hemisphere. This drift was first described in 1596 by the Dutch map maker Abraham Ortelius and later in 1858 by a French geographer, Antonio Snider-Pellegrini, who based his theory not only on the shape of the continents on the world map (Fig. 8.1) but also on the evidence provided by plant fossils found in both Europe and the United States, which were identical (Pellegrini 2019). However, Alfred Wegener was the person who effectively spread the message. He was intrigued by the striking similarity between the eastern coastline of South America and the western coastline of Africa (Wegener 1920). In Chap. 1 of his book, we read (Wegener 1966):

> *(…) The first concept of continental drift first came to me as far back as 1910, when considering the map of the world, under the first impression produced by the congruence of the coastlines on either side of the Atlantic.*

The continental drift theory proposed by Alfred Wegener, although rejected and forgotten for decades, commenced another revolution in the Earth sciences (Petrović and Marković 2012; Rundić 2012). The theory slowly evolved into the concept of plate tectonics, which explains the mechanisms underlying the formation and disappearance of the oceanic lithosphere and movements of the Earth's crust. Three mechanisms and plate boundaries are fundamental to global plate tectonics (Stern 2018):

1. Formation of new oceanic plate material by **seafloor spreading** at mid-ocean ridges (divergent boundaries) (Fig. 8.2).
2. Destruction of old oceanic lithosphere that sinks at **subduction** zones (convergent boundaries).
3. Sliding of two plates past each other along **transform faults** (transform boundary).

Diffuse plate boundaries are zones of deformation that occur in the continental and oceanic lithosphere and can be hundreds to thousands of kilometers wide (Zatman et al. 2005). The velocities of movements along plate boundaries differ and can now be monitored using a global positioning system (GPS).

Several examples of events connected to tectonic plate movements and volcanic eruptions are known from recent Earth history. However, on the basis of their damage and death toll, the 2004 tsunami in the Indian Ocean has been the most impactful (Ioualalen et al. 2007). The event resulted in >292,000 fatalities in 12 countries, mostly along the western coast of Sumatra (Gusiakov et al. 2019). As shown in Fig. 8.2, along the western coast of Indonesia, there is a convergent (subduction) zone between the Sunda (formerly Eurasian), Australian and Indian plates. The tsunami was induced by an oceanic bottom earthquake, with the epicenter 80 km west of the coast of northern Sumatra; it was induced following strain accumulation at the Indian/Sunda junction since the last large earthquake that occurred there in 1861 (Ioualalen et al. 2007). The Indian and Sunda plates move on the order of 4 cm per year in the direction of 20°N, and the Australian and Sunda plates move on the order of 5 cm per year in the direction of 8°N (Ioualalen et al. 2007).

8.1 Description of the Data and Project Outline

There are several online resources that provide spatial data in various formats. The web service of the United States Geological Survey (USGS) offers access to data on plate boundaries and the velocity of plate movement (https://www.usgs.gov/faqs/how-fast-do-tectonic-plates-move). The data can be downloaded in KMZ format as a zipped KML (keyhole markup language) file. KML is a format used to display spatial data in services such as Google Earth, Google Maps, and Big Maps. There are GitHub repositories that offer access to published data. For example, the tectonic plate repository (https://github.com/fraxen/tectonicplates/tree/master) includes data on digital models of plate boundaries (Bird 2003). More recent data are available from the global-tectonic repository in GitHub (https://github.com/dhasterok/global_tectonics/tree/main) or Zenodo (https://zenodo.org/records/5093930) (Hasterok et al. 2022). There are also R packages with incorporated datasets on global tectonics and reconstructions of their movements (Müller et al. 2018; Stephan et al. 2023; Kocsis et al. 2024).

Historical earthquake data can be obtained from several online resources. For example, the National Centers for Environmental Information (NCEI) of the NOAA provides access to earthquake strong motion data (1933–1994), with over 15,000 records (https://www.ngdc.noaa.gov/hazard/strong.shtml). For tasks in this project, we use built-in earthquake datasets from the **geostats** (Vermeesch 2023) and **MPV** (Braun and MacQueen 2024) packages.

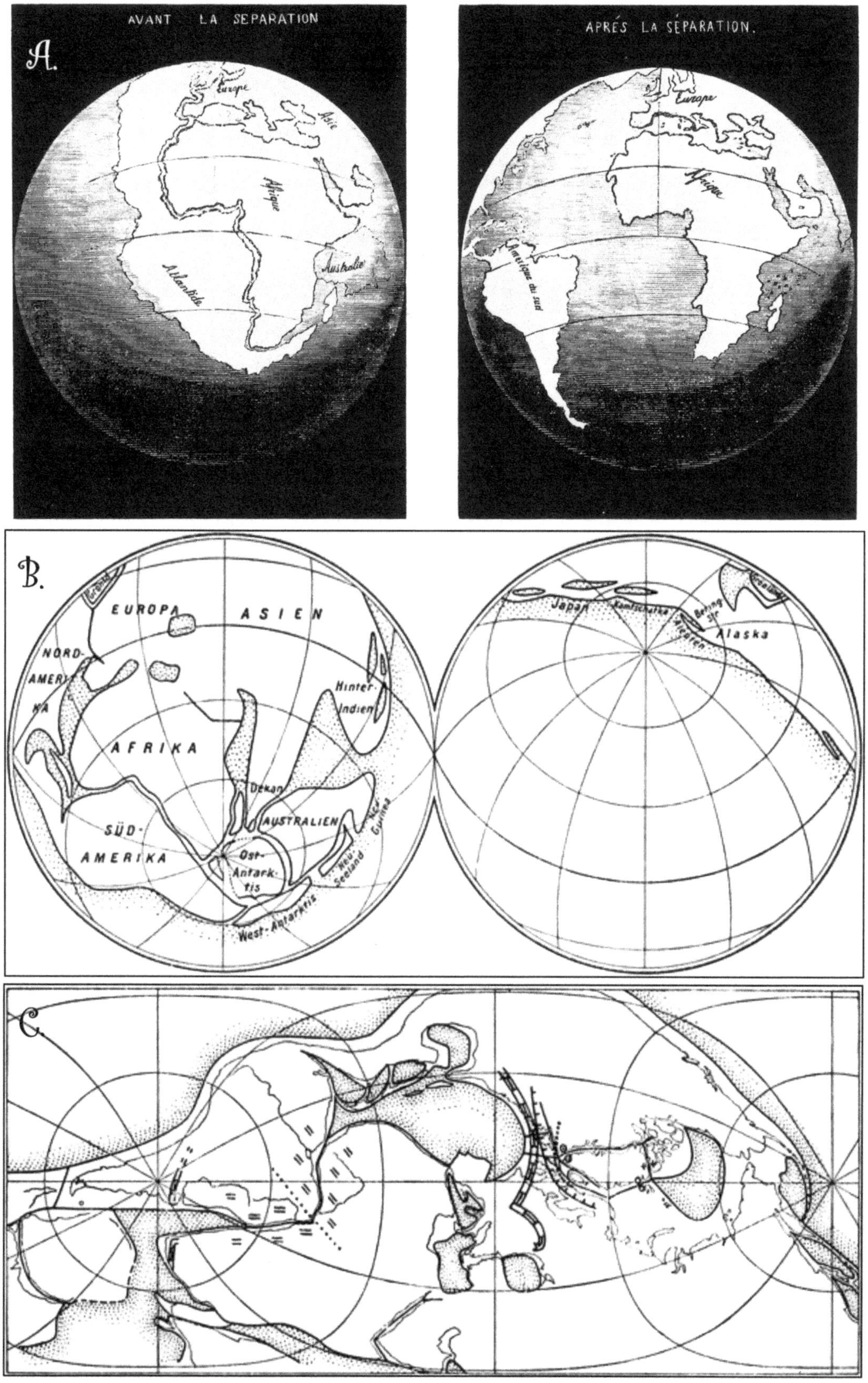

Fig. 8.1 Three maps illustrating early attempts at explaining continental drift theory. (**a**) Antonio Snider-Pellegrini's maps from 1858 (Public Domain, https://commons.wikimedia.org/w/index.php?curid=400158). (**b** and **c**) are from Alfred Wegener's book, which illustrates the concept of changes in the geographical position of lands. (**b**) Position of continental plates in the Carboniferous period (without consideration of the

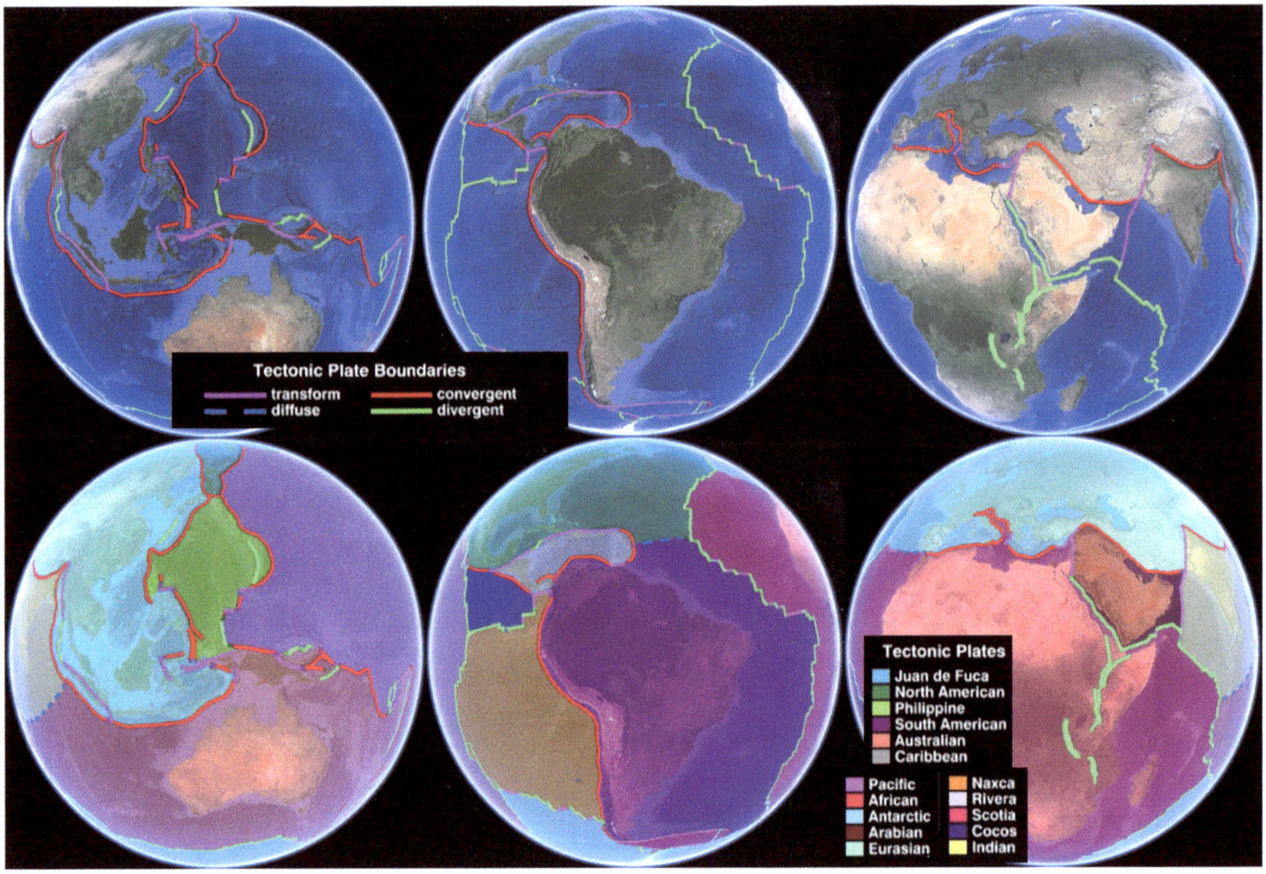

Fig. 8.2 Tectonic plates and their boundaries. Data source: United States Geological Survey's Earthquake Hazards Program (https://svs.gsfc.nasa.gov/vis/a000000/a002900/a002953/a002953.kml). The compilation was prepared using background maps from Google Earth Pro

8.2 R in Practice: Global Tectonics Data Synthesis and Interpretation

In this project, we use open data that are available from the internet or are an integral part of dedicated R packages. Our main objectives are static and dynamic (as an animation) data visualization. Two aspects of plate tectonics are examined: 1. plate tectonics and their boundaries and 2. past transformations (movements and transformations) of lands and oceans. Another task is to explore earthquake dynamics at the spatiotemporal scale. The first map is a mosaic of tectonic plate boundaries and coastlines (Fig. 8.3). The plate boundary data are from Bird (2003) and can be used in R after the **tectonicr** package is installed (Stephan et al. 2023). The first layer in Fig. 8.3 contains continent coastlines, which constitute a dataset within the **rnaturalearth** package (Massicotte and South 2023). The second layer contains tectonic plate boundaries. Both objects are of the simple feature (***sf***) class. The ***sf*** class represents spatial data in R and contains geometry-defining positions for ***sf*** objects on the Earth's surface (https://r-spatial.github.io/sf/articles/sf1.html). The geometric features can be points, lines, or polygons. An ***sf*** object can also contain attributes stored in a data frame. The attributes are properties of elements (qualitative or quantitative) and are related to the object geometry type and coordinates. The divergent plate boundary, shown in Fig. 8.3, extends through the central part of the Atlantic Ocean from south to north. It is a spreading zone where new oceanic crust forms in the mid-Atlantic ridge system at a rate from 2 to 5 centimeters per year. Note that the map is centered on the Atlantic Ocean. This perspective can be changed to the Pacific Ocean using the ***st_shift_longitude()*** function from the **sf** package (Pebesma 2018) or the ***st_wrap_x()***

Fig. 8.1 (Continued) water cover) (original figure caption: "*Lage der Kontinentalschollen für die Karbonzeit (ohne Rücksicht auf Wasserbedeckung)*"), (**c**) Position of the Atlantic continental plates in the Eocene (without consideration of the water cover) (original figure caption: "*Lage der atlantischen Kontinentalschollen im Eozän (ohne Rücksicht auf die Wasserverteilung)*). Source: The Project Gutenberg (www.gutenberg.org)

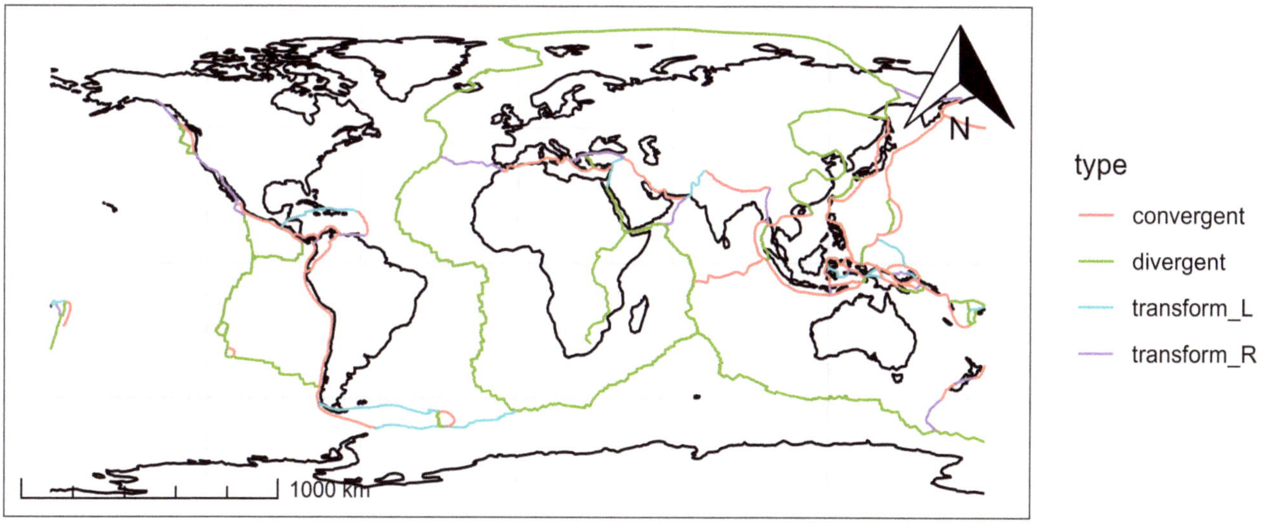

Fig. 8.3 Tectonic plate boundaries (Bird 2003)

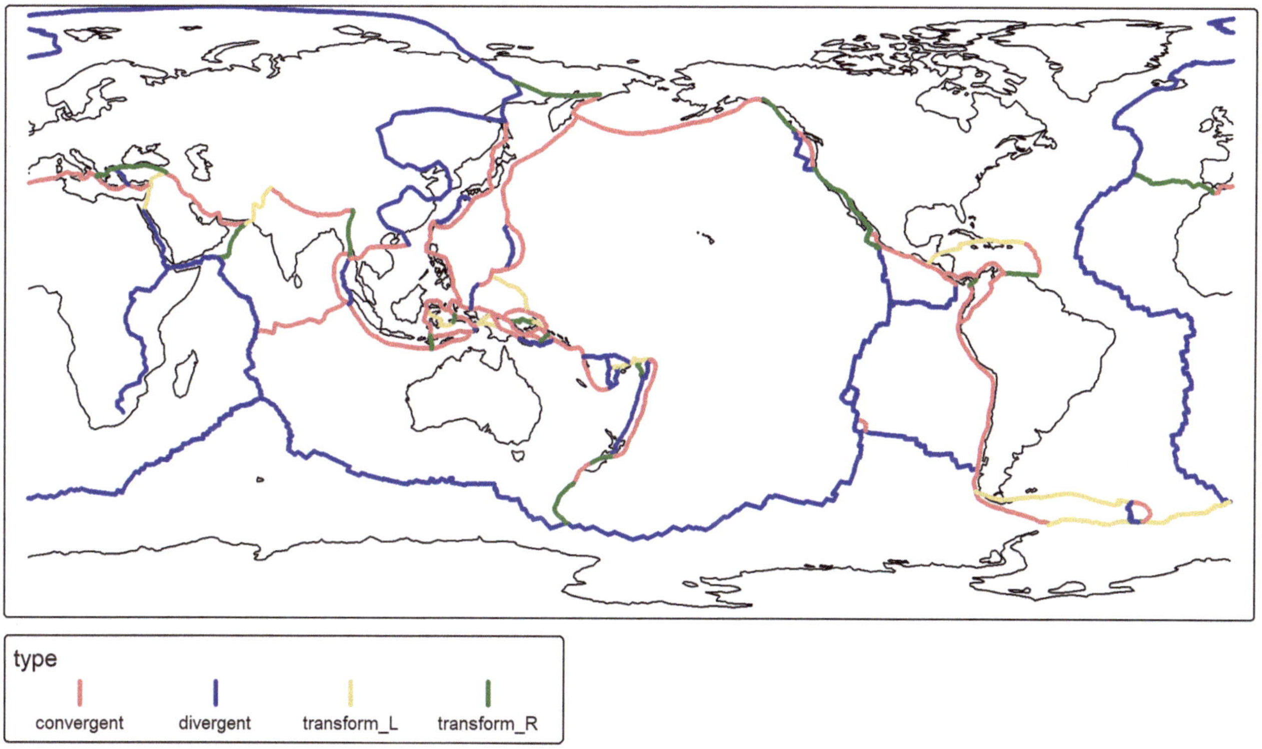

Fig. 8.4 The map displays the same data as in Fig. 8.3, but this map was centered on the Pacific Ocean

function from the **lwgeom** package (Pebesma 2024). This way of displaying global spatial data is better if, for instance, the Ring of Fire surrounding the Pacific Ocean is to be shown (Fig. 8.4). This zone is characterized by extensive volcanic and earthquake activity (Figs. 8.5 and 8.6).

```
pacman::p_load("tectonicr", "rnaturalearth",
"ggspatial", "sf", "rgplates", "tmap", "MPV",
"lwgeom")
```

```
library(tectonicr) #plate motion modelling
library(rnaturalearth) #access to spatial data
library(ggspatial) #spatial data and
visualization using ggplot2
library(sf) #spatial vector data analyses
library(ggplot2) #data plotting and
visualization

#We start by importing the coastline dataset.
```

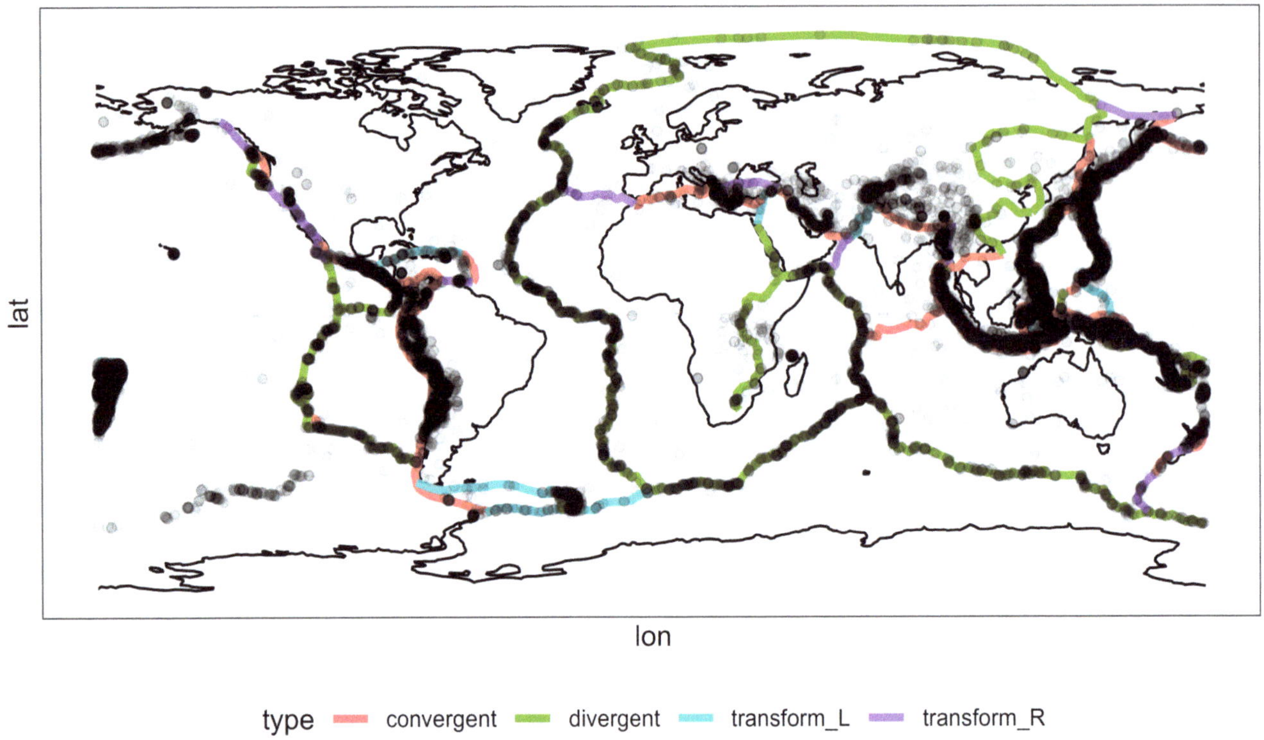

Fig. 8.5 The global distribution of earthquakes from 2017 to 2020 and tectonic plate boundaries. The map was prepared with the **ggplot2** package

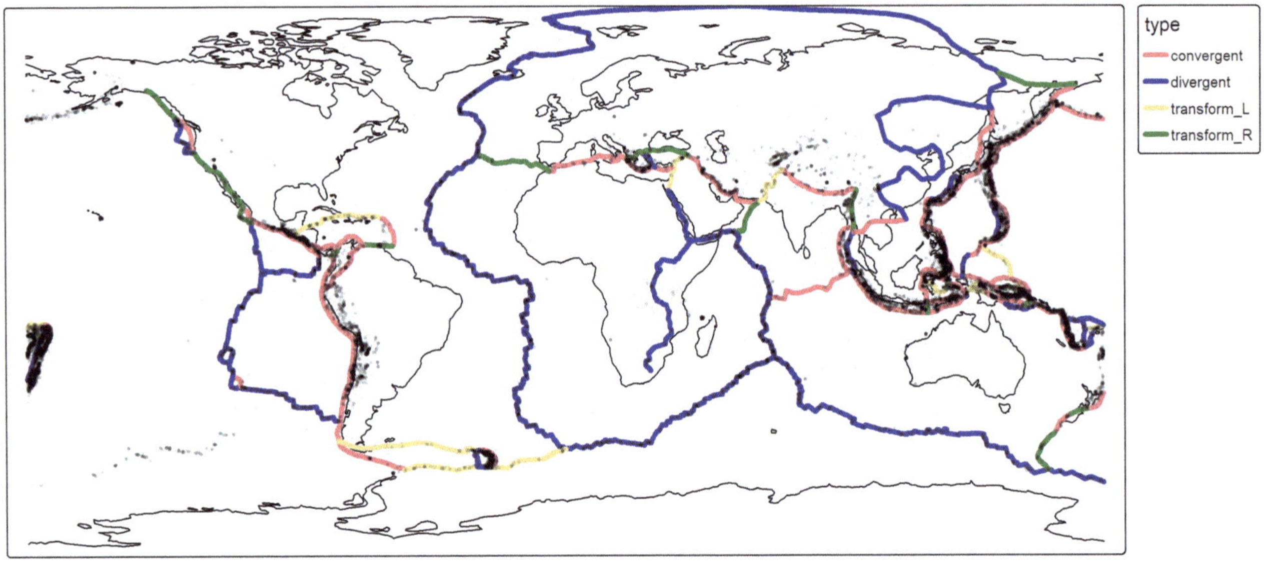

Fig. 8.6 The data displayed on the map are the same as those in Fig. 8.5. The map was prepared with the **tmap** package

```
#The ne_coastline() function comes from the
rnaturalearth package.

coast <- ne_coastline()#object of class 'sf'

#Now we load plate boundaries according to Bird
(2003). The dataset is a part of the tectonicr
package.

data("plates", package = "tectonicr")

#What is the geometry type of this spatial
dataset?
unique(st_geometry_type(plates))

#We change it to lines.
boundaries <- st_cast(plates, "LINESTRING")
```

```
#Plotting the data.
ggplot()+
 geom_sf(data = coast)+
 geom_sf(data = boundaries, mapping = aes(color
 = type))+
 theme_bw()+
 annotation_scale(location = "bl", style =
 "ticks")+
 annotation_north_arrow(location = "tr", which_
 north = "true")

#Saving the plot.
#ggsave("plates_bird2003.png", dpi=600)

#Changing the perspective to the Pacific Ocean.

coast_pac1 <- lwgeom::st_wrap_x(coast, 0,360)
plates_pac1 <- lwgeom::st_wrap_x(plates, 0,360)

#Note the difference in the longitudinal range
(x_min, x_max) of the coast object (centered on
the Atlantic Ocean) and the coast_pac1 object
(centered on the Pacific Ocean).

st_bbox(coast)
st_bbox(coast_pac1)

#Plot a map with transformed coordinates.
pac <- tm_shape(coast_pac1)+
  tm_lines()+
  tm_shape(plates_pac1)+
  tm_lines(col = "type", lwd = 3)+
  tm_legend(orientation = "landscape")

pac

#tmap_save(pac, "plates_bird2003_pacific.png")

#Use the earthquake data from the geostats
package and plot on the map of the tectonic
plates.

pacman::p_load("geostats", "lubridate",
"ggpubr")

library(geostats)

data(earthquakes, package='geostats')


#Plot it using the ggplot2 package.
ggplot()+
 geom_sf(data = coast)+
 geom_sf(data = boundaries, mapping = aes(color
 = type), linewidth = 1.5)+
 geom_point(earthquakes, mapping = aes(x=lon,
 y=lat), alpha = 0.1)+
 theme_bw()+
 theme
(legend.position = "bottom")

#Saving the plot.
#ggsave("plates_bird2003_and_earthquakes.png",
dpi=600)

#Plot it using the tmap package.
#First, we need to create the sf object using
the data frame with the earthquake data. When
defining the crs, we use the EPSG code for
WGS84, which is 4326.

eq_sf <- st_as_sf(earthquakes, coords =
c("lon", "lat"), crs = 4326)

library(tmap)

bird2003 <- tm_shape(coast)+
 tm_lines()+
 tm_shape(boundaries)+
 tm_lines(col = "type", lwd = 4)+
 tm_shape(eq_sf)+
 tm_dots(fill_alpha = 0.1)

bird2003

#Saving the map.
#tmap_save(bird2003, "plates_bird2003_and_
earthquakes_tmap.png")
```

We can obtain more information from this dataset. For example, we can plot time series and periods with relatively high earthquake intensities (magnitudes). In this way, we can display the data differently by transferring them from the spatial domain to the temporal domain. However, plots of the entire dataset can be difficult to interpret; therefore, it is sometimes good to focus on extreme values that characterize better the analyzed system (Fig. 8.7). We can further reduce the size of the dataset and explore only a selected region or country. In this example, we focus on Japan and a 20 km buffer zone from the coastline. We show the time series in the upper panel, and the spatial redistribution of earthquakes is illustrated with four classes indicating the event year (Fig. 8.8).

```
#Plot the time series and periods with higher
earthquake intensity (magnitude).

#First step: creating a column with earthquake
time occurrences.
```

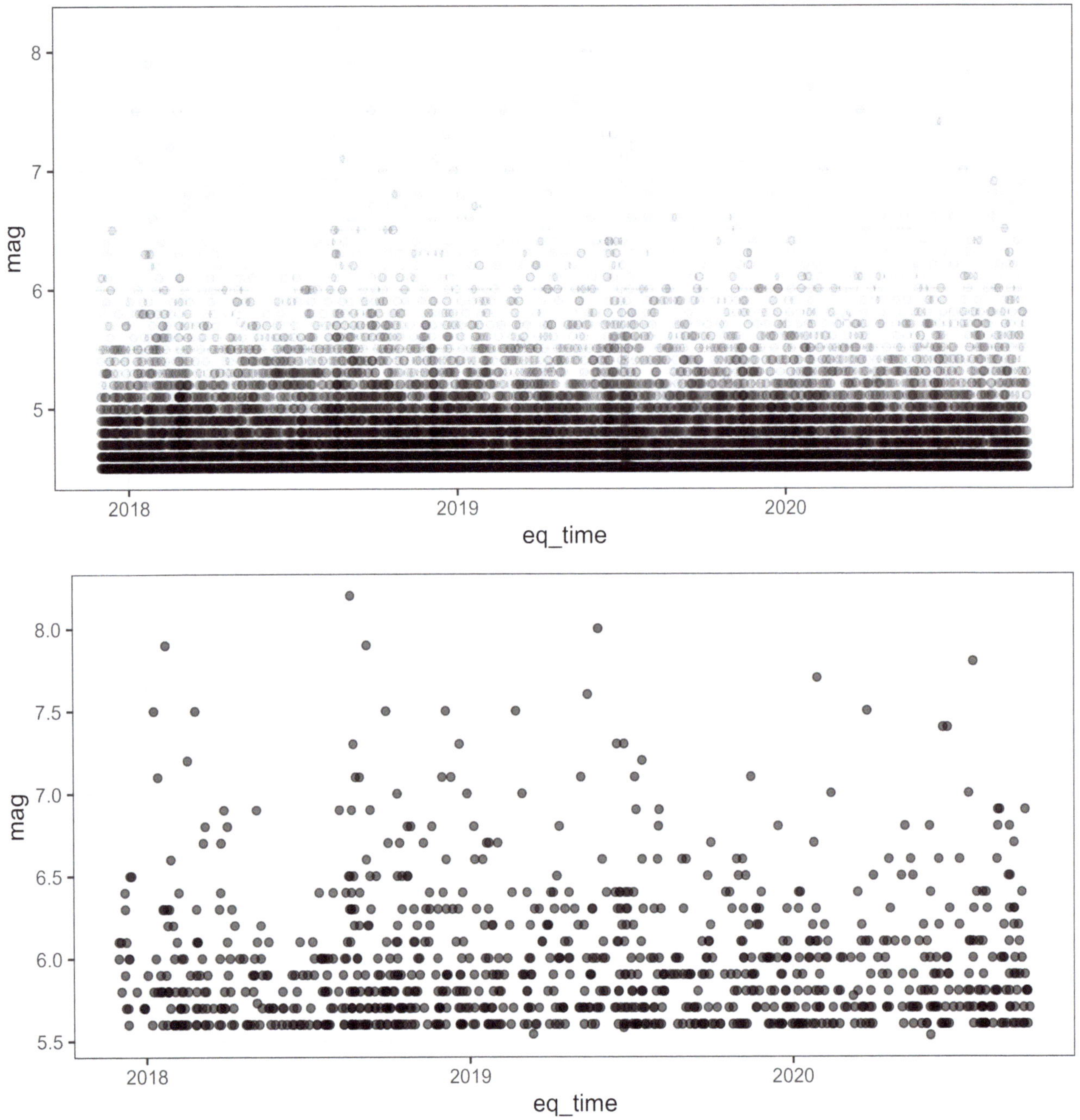

Fig. 8.7 The time series showing earthquake magnitudes for the full dataset (upper panel) and for earthquakes with magnitudes greater than 5.5 (lower panel)

```
library(lubridate)

#Time of earthquakes.
eq1 <- earthquakes |>
 mutate(eq_time = ymd_hms(paste(year, month,
 day, hour, minute, second, sep = "-")))

#Plot the data.
pl1 <- ggplot()+
  geom_point(eq1, mapping = aes(x=eq_time,
  y=mag), alpha = 0.1)+
  theme_bw()

#Filter and plot the data.
pl2 <- ggplot(eq1[eq1$mag>5.5,], mapping =
aes(x=eq_time, y=mag))+
  geom_point(aes(alpha = 0.1))+
  theme_bw()+
  theme
```

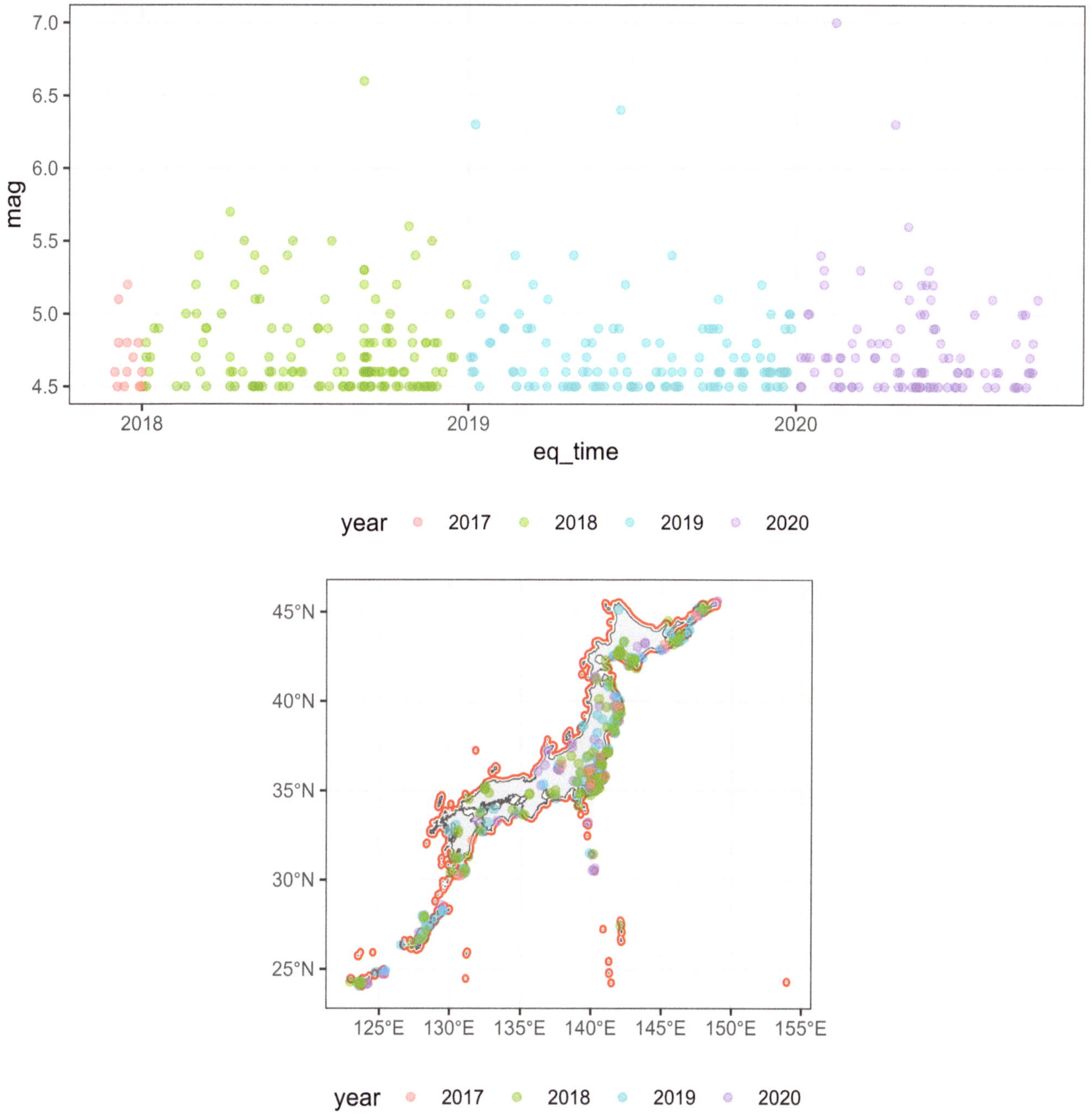

Fig. 8.8 The earthquake dataset was reduced to the Japanese territory and a 20 km buffer zone from the coastline. Points were categorized according to the year of earthquake occurrence

```
(legend.position = "none")

library(ggpubr)

ggarrange(pl1, pl2, nrow = 2)

#Save the plot.
#ggsave("eq_mag.png", dpi = 600)
```

```
#We will filter the data for Japan and show the
progress of earthquake activity in this
country.

library(rgeoboundaries)

#Get the spatial data for Japan.
japan <- gb_adm0("Japan")
```

```
#Calculate the buffer zone of 20 km from the
coastline. After the buffer zone operation, we
obtain the SpatVector object (the result of the
terra::buffer() function), which is then
converted to the sf object using the st_as_sf()
function. We do that because we want to display
two sf objects in one plot.

japan_buff <- st_as_sf(terra::buffer(terra::vect(
japan), 20000)) #20 km distance given in meters

#Inspect the data.
plot(japan_buff$geometry, col="green")
plot(japan$geometry, add=TRUE)

#Spatial join operation allows for filtering
earthquakes that occurred within the Japan
area, including the 20 km buffer zone.
jap_eq <- st_join(x=eq_sf, y=japan_buff, join =
st_within, left = FALSE)

#Show the time series and map on one plot.
jap_eq_date <- jap_eq %>%
 mutate(eq_time = ymd_hms(paste(year, month,
 day, hour, minute, second, sep = "-")))

#Time series.
jap_eq_ts <- ggplot()+
 geom_point(jap_eq_date, mapping = aes(x=eq_
 time, y=mag, color=as.factor(year)), alpha =
 0.5)+
 theme_bw()+
 guides(col = guide_legend("year"))+
 theme
(legend.position = "bottom")

jap_eq_map <- ggplot()+
 geom_sf(data = japan)+
 geom_sf(data = st_cast(st_as_sf(japan_buff),
 "MULTILINESTRING"), color = "red")+
 geom_sf(data = jap_eq, aes(color =
 as.factor(year)), alpha = 0.5)+
  theme_bw()+
  guides(col = guide_legend("year"))+
  theme
(legend.position = "bottom")

ggarrange(jap_eq_ts, jap_eq_map, nrow = 2)

#Saving the plot.
#ggsave("japan_ts_map.png", dpi = 600)
```

Our next task is to determine how many earthquakes occurred along a certain tectonic plate boundary type and within its 100 km wide buffer zone. The dataset of 20,000 earthquakes between 2017 and 2020 was derived from the USGS Earth Hazard Program web service (https://earthquake.usgs.gov/earthquakes/search/). It is a part of the **geostats** package, so it can be called directly to the session memory with the ***data()*** function. Each type of plate boundary is filtered from the Bird dataset (Bird 2003), and the buffer zone is calculated. A spatial join is then applied through the ***st_join()*** function to select only earthquake points within the buffer. Over 50% of the earthquakes in the 2017-2020 period occurred along convergent plate boundaries (Table 8.1). This type of plate boundary is dominant around the Pacific Ocean (Figs. 8.4 and 8.9).

Another interesting fact we can learn from earthquake data implemented in the **MPV** package (Braun and MacQueen 2024). It is the relationship between magnitude and focal depth. Considering only earthquakes with magnitudes greater than 5.8 from the period of 1964-1985, high-magnitude earthquakes tend to occur close to ocean bottoms or land surfaces (Fig. 8.10).

```
#Next, we will count the number of earthquakes
along the plate boundaries of each class:
convergent, divergent, transform_L,
transform_R. First, a 100 km buffer is
calculated around different tectonic plate
types. Second, the st_join() function extracts
only the points within each buffer zone. This
operation is a type of spatial selection
algorithm.

#Divergent plate boundary.
div <- boundaries[boundaries$type ==
"divergent",]
div_buff <- terra::buffer(terra::vect(div),
100000)

#Convergent plate boundary.
conv <- boundaries[boundaries$type ==
"convergent",]
conv_buff <- terra::buffer(terra::vect(conv),
100000)

#Transform_L plate boundary.
```

Table 8.1 Number of points along tectonic plate boundaries and within their buffer zones, which range to a distance of 50 km from each side of the boundary line

Type	*n*	Percent
convergent	9323	58.7
divergent	4313	27.1
transform_R	988	6.2
transform_L	1267	8
total *n*	15891	100

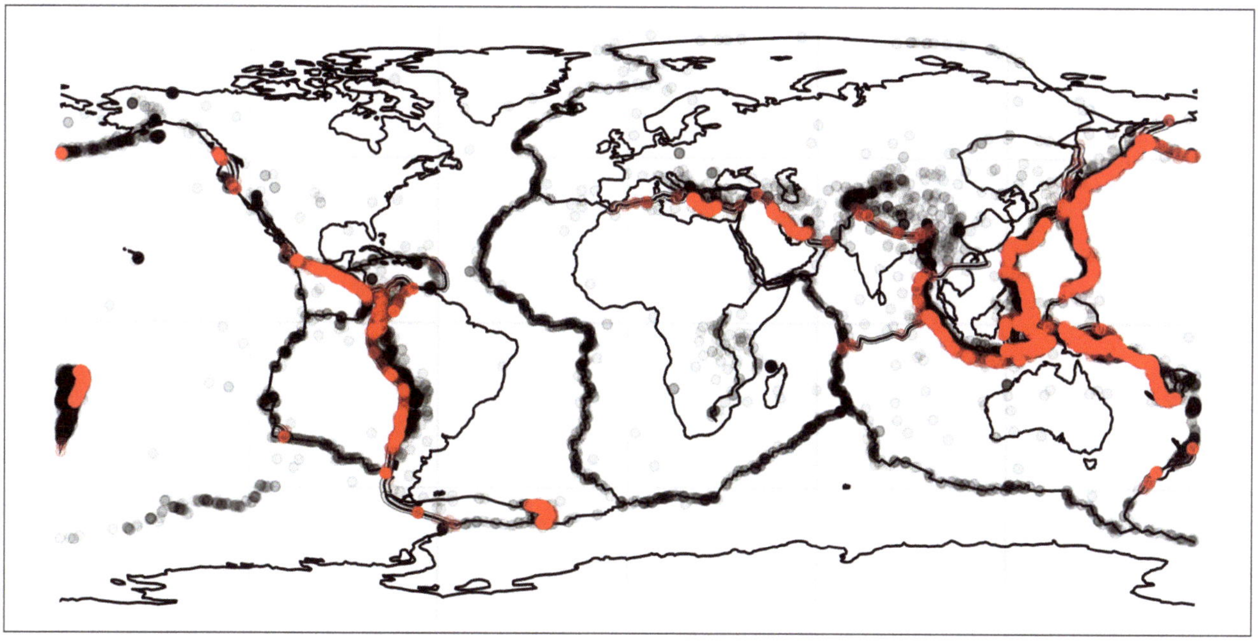

Fig. 8.9 Earthquakes recorded in the 2017-2020 period (black dots) and earthquake activity along convergent plate boundaries (red dots) for the same period

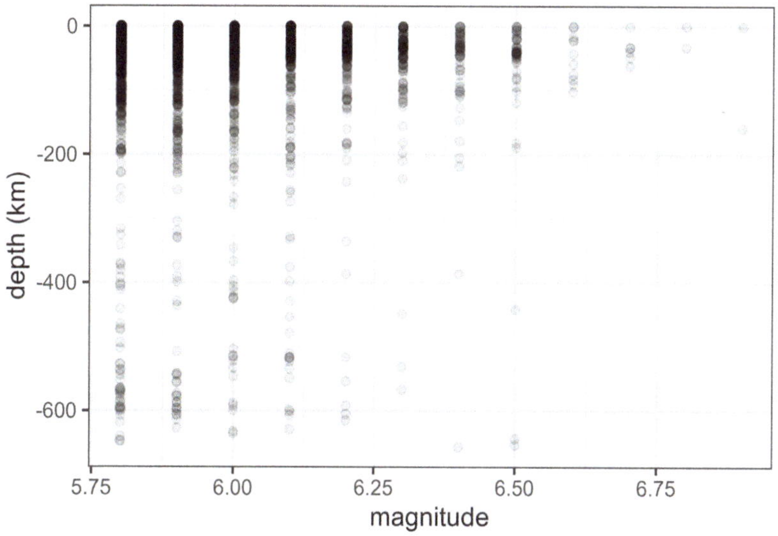

Fig. 8.10 Relationships between earthquake magnitudes greater than 5.8 and focal depths based on records from 1964 and 1985. The dataset is a part of the MPV package (Braun and MacQueen 2024)

```
tl <- boundaries[boundaries$type ==
"transform_L",]
tl_buff <- terra::buffer(terra::vect(tl), 100000)

#Transform_R plate boundary.
tr <- boundaries[boundaries$type ==
"transform_R",]
tr_buff <- terra::buffer(terra::vect(tr), 100000)

#Spatial selection.
conv_buff_eq <- st_join(x=eq_sf, y=st_as_
sf(conv_buff), join = st_within, left = FALSE)
div_buff_eq <- st_join(x=eq_sf, y=st_as_sf(div_
buff), join = st_within, left = FALSE)
tl_buff_eq <- st_join(x=eq_sf, y=st_as_sf(tl_
buff), join = st_within, left = FALSE)
tr_buff_eq <- st_join(x=eq_sf, y=st_as_sf(tr_
buff), join = st_within, left = FALSE)

library(gt)
```

```
#Build a table.
eq_sum <- data.frame(type = c(unique(conv_buff_
eq$type),
                unique(div_buff_eq$type),
                unique(tr_buff_eq$type),
                unique(tl_buff_eq$type)),
                n = c(length(conv_buff_eq$type),
                length(div_buff_eq$type),
                length(tr_buff_eq$type),
                length(tl_buff_eq$type)))

#Calculate the total number of points and add a
new row to the table.
eq_sum1 <- eq_sum |>
 dplyr::mutate(percent = round((n*100)/
 sum(n),1)) |>
 rbind(c(type = "total n", n = sum(eq_sum$n),
 percent = sum((eq_sum$n*100)/sum(eq_sum$n))))
 |>
 gt()

eq_sum1

#Save the table.
#gtsave(eq_sum1, "eq_gt_table.png")

#Plot the data with extracted points along the
convergent plate boundaries.
ggplot()+
  geom_sf(data = coast)+
  geom_sf(data = st_as_sf(conv_buff))+
  geom_sf(data = boundaries)+
  geom_sf(data = eq_sf, alpha = 0.1)+
  geom_sf(data = conv_buff_eq, color = "red",
  alpha = 0.1)+
  theme_bw()

#Save the map.
#ggsave("conv_eq.png", dpi = 600)

#Using the earthquake data from the MPV
package, we can test the relationship between
the magnitude and focal depth (the depth at
which an earthquake occurs).

library(MPV)
data(earthquake, package='MPV')

names(earthquake)#variable (column) names

#Scatterplot.
ggplot()+
  geom_point(earthquake, mapping = aes(x =
  magnitude, y = -depth), alpha = 0.1)+
  theme_bw()+
  ylab("depth (km)")

#Save the plot.
#ggsave("depth_magn.png", dpi = 600, width =
12, height = 8, units = "cm")
```

The last task in this chapter is related to one of the key problems that has been unsolved for several decades since early geological surveys were first performed. This problem involves the evolution and formation of continents and ocean bottoms. Reconstruction of the paleopositions of continents has been the subject of several scientific projects (Golonka et al. 2006; Matthews et al. 2016; Scotese 2016; Torsvik and Cocks). The systematic information from these projects in the form of spatial data that can be used for visualization is available through the **rgplates** package (Müller et al. 2018; Kocsis et al. 2024). The package is an R interface for the Gplates Web Service (https://gwsdoc.gplates.org/) and the Gplates Desktop Application (https://www.gplates.org/). For data visualization and animation, we use the ***reconstruct()*** function from the **rgplates** package. The "MERDITH2021" model is set as the default for the model argument of the ***reconstruct()*** function. It means the function will use paleogeographical reconstructions developed by Merdith et al. (2021). Figure 8.11 shows the reconstruction for nine periods from 350 Ma to 1 Ma. This figure was prepared using the **tmap** package.

```
library(rgplates)

#Rotation of plates using the GPlates Web
Service
coastlines350 <- reconstruct("coastlines",
age=350)
coastlines300 <- reconstruct("coastlines",
age=300)
coastlines250 <- reconstruct("coastlines",
age=250)
coastlines200 <- reconstruct("coastlines",
age=200)
coastlines150 <- reconstruct("coastlines",
age=150)
coastlines100 <- reconstruct("coastlines",
age=100)
coastlines50 <- reconstruct("coastlines",
age=50)
coastlines20 <- reconstruct("coastlines",
age=20)
coastlines1 <- reconstruct("coastlines", age=1)

#Set the edge of the map (for mollweide).
edge <- mapedge()
```

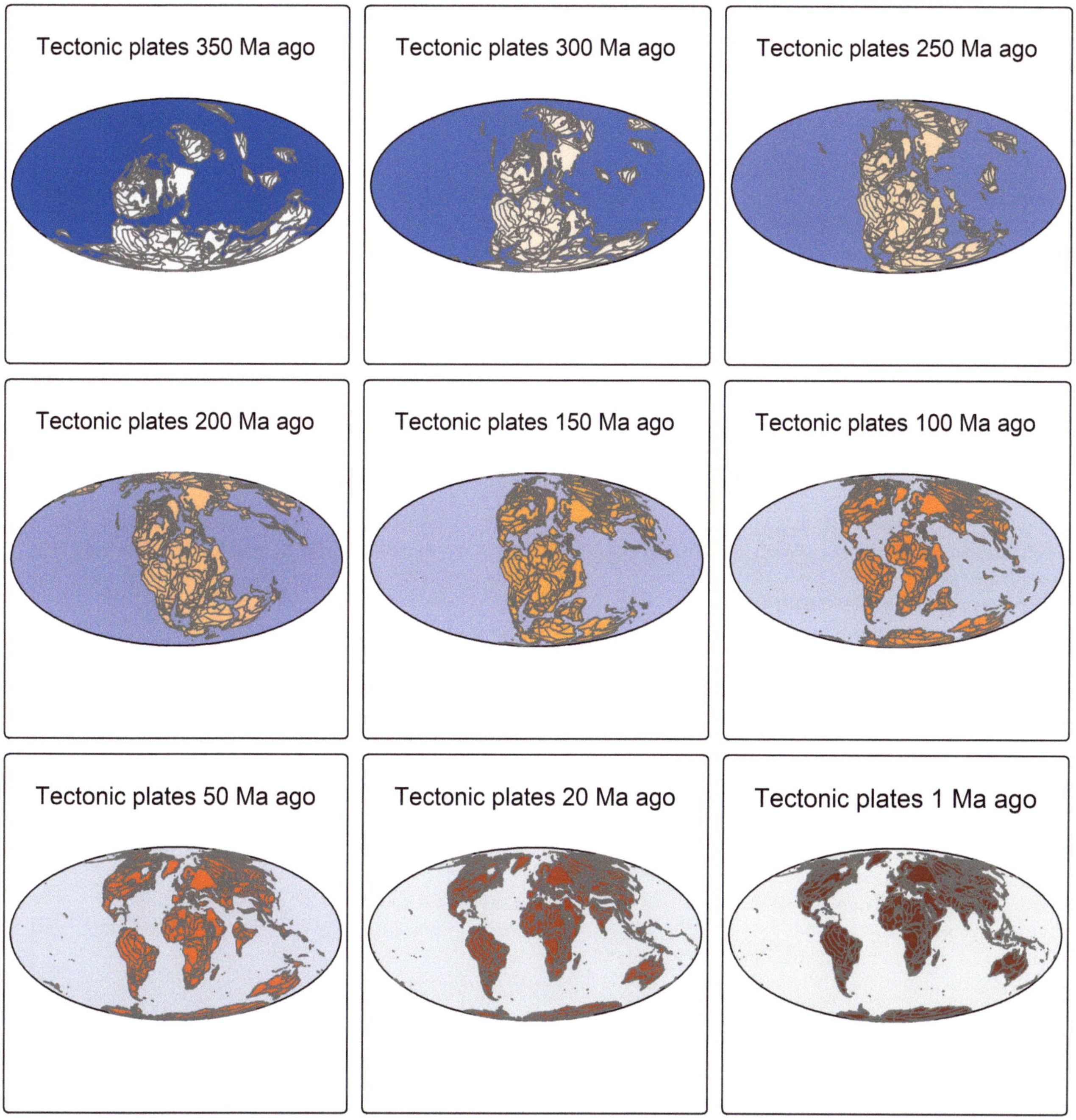

Fig. 8.11 Maps showing the geographical positions of continents during nine periods from 350 Ma to 1 Ma. We used the default model for the ***reconstruct()*** function in the **rgplates** package developed by Merdith et al. (2021)

```
#Transform to Mollweide projection. It is also
known as elliptical projection. It allows the
display of the Earth's surface on the map with
minimal distortions around polar regions.

epsg <- "ESRI:54009"
edgeMoll <- sf::st_transform(edge, crs=epsg)
coastsMoll350 <- sf::st_
transform(coastlines350, crs=epsg)
coastsMoll300 <- sf::st_
transform(coastlines300, crs=epsg)
coastsMoll250 <- sf::st_
transform(coastlines250, crs=epsg)
coastsMoll200 <- sf::st_
transform(coastlines200, crs=epsg)
```

```
coastsMoll150 <- sf::st_
transform(coastlines150, crs=epsg)
coastsMoll100 <- sf::st_
transform(coastlines100, crs=epsg)
coastsMoll50 <- sf::st_transform(coastlines50,
crs=epsg)
coastsMoll20 <- sf::st_transform(coastlines20,
crs=epsg)
coastsMoll1 <- sf::st_transform(coastlines1,
crs=epsg)

#Select color scales for the graphical display
of the data. The hcl.colors() creates a vector
of n contiguous colors.
grDevices::hcl.colors(9, "Blues 2")
grDevices::hcl.colors(9, "Oranges")

library(tmap)

map350 <- tm_shape(edgeMoll)+
 tm_polygons(fill = "#023FA5", col = "black")+
 tm_shape(coastlines350)+
 tm_polygons(fill="#FEF5EC")+
 tm_title_in("Tectonic plates 350 Ma ago")

map300 <- tm_shape(edgeMoll)+
 tm_polygons(fill = "#485CA8", col = "black")+
 tm_shape(coastlines300)+
 tm_polygons(fill="#FFE5CC")+
  tm_title_in("Tectonic plates 300 Ma ago")

map250 <- tm_shape(edgeMoll)+
  tm_polygons(fill = "#6A76B2", col = "black")+
  tm_shape(coastlines250)+
  tm_polygons(fill="#FFCDA1")+
  tm_title_in("Tectonic plates 250 Ma ago")

map200 <- tm_shape(edgeMoll)+
  tm_polygons(fill = "#878FBD", col = "black")+
  tm_shape(coastlines200)+
  tm_polygons(fill="#FDB16E")+
  tm_title_in("Tectonic plates 200 Ma ago")

map150 <- tm_shape(edgeMoll)+
  tm_polygons(fill = "#A1A6C8", col = "black")+
  tm_shape(coastlines150)+
  tm_polygons(fill="#F7921E")+
  tm_title_in("Tectonic plates 150 Ma ago")

map100 <- tm_shape(edgeMoll)+
  tm_polygons(fill = "#B7BBD1", col = "black")+
  tm_shape(coastlines100)+
  tm_polygons(fill="#ED6F00")+
  tm_title_in("Tectonic plates 100 Ma ago")

map50 <- tm_shape(edgeMoll)+
  tm_polygons(fill = "#CBCDD9", col = "black")+
  tm_shape(coastlines50)+
  tm_polygons(fill="#D35000")+
  tm_title_in("Tectonic plates 50 Ma ago")

map20 <- tm_shape(edgeMoll)+
  tm_polygons(fill = "#DADBDF", col = "black")+
  tm_shape(coastlines20)+
  tm_polygons(fill="#AA3C00")+
  tm_title_in("Tectonic plates 20 Ma ago")

map1 <- tm_shape(edgeMoll)+
  tm_polygons(fill = "#E2E2E2", col = "black")+
  tm_shape(coastlines1)+
  tm_polygons(fill="#802A07")+
  tm_title_in("Tectonic plates 1 Ma ago")

#Create animation.
tmap_animation(list(map350, map300, map250,
map200, map150, map100, map50, map20, map1),
width = 600, height = 500, delay = 80)

#The animation can be saved as a gif file or a
video file, if the filename argument is specified.


#The maps can also be arranged in one figure and
saved as a PNG file.
#plate_anim <- tmap_arrange(map350, map300,
map250, map200, map150, map100, map50, map20,
map1, ncol=3, nrow=3)

#Save the figure.
#tmap_save(plate_anim, "plates_anim.png", dpi =
600)
```

References

Bird P (2003) An updated digital model of plate boundaries. Geochem Geophys Geosyst 4:1. https://doi.org/10.1029/2001GC000252

Braun WJ, MacQueen S (2024) MPV: data sets from Montgomery. Peck and Vinning

Golonka J, Krobicki M, Pająk J et al (2006) Global plate tectonics and paleogeography of Southeast Asia. AGH University of Science and Technology, Kraków

Gudmundsson MT, Larsen G, Höskuldsson Á, Gylfason ÁG (2008) Volcanic hazards in Iceland. Jökull 58:251–268

Gusiakov VK, Dunbar PK, Arcos N (2019) Twenty-five years (1992-2016) of global tsunamis: statistical and analytical overview. Pure Appl Geophys 176:2795–2807. https://doi.org/10.1007/s00024-019-02113-7

Hasterok D, Halpin JA, Collins AS et al (2022) New maps of global geological provinces and tectonic plates. Earth Sci Rev 231:104069. https://doi.org/10.1016/j.earscirev.2022.104069

Ioualalen M, Asavanant J, Kaewbanjak N et al (2007) Modeling the 26 December 2004 Indian Ocean tsunami: case study of impact in Thailand. J Geosphys Res Oceans 112:C07024. https://doi.org/10.1029/2006JC003850

Kocsis AT, Raja NB, Williams S (2024) rgplates: R interface for the GPlates web service and desktop application

Massicotte P, South A (2023) rnaturalearth: world map data from natural earth

Matthews KJ, Maloney KT, Zahirovic S et al (2016) Global plate boundary evolution and kinematics since the late Paleozoic. Glob Planet Chang 146:226–250. https://doi.org/10.1016/j.gloplacha.2016.10.002

Merdith AS, Williams SE, Collins AS et al (2021) Extending full-plate tectonic models into deep time: linking the Neoproterozoic and the Phanerozoic. Earth Sci Rev 214:103477. https://doi.org/10.1016/j.earscirev.2020.103477

Müller RD, Cannon J, Qin X et al (2018) GPlates: building a virtual earth through deep time. Geochem Geophys Geosyst 19:2243–2261. https://doi.org/10.1029/2018GC007584

Niu Y (2014) Geological understanding of plate tectonics: basic concepts, illustrations, examples and new perspectives. Global Tectonics and Metallogeny 10:23. https://doi.org/10.1127/gtm/2014/0009

Pebesma E (2018) Simple features for R: standardized support for spatial vector data. R J 10:439–446. https://doi.org/10.32614/RJ-2018-009

Pebesma E (2024) Lwgeom: bindings to selected "liblwgeom" functions for simple features

Pellegrini PA (2019) Styles of thought on the continental drift debate. J Gen Philos Sci 50:85–102. https://doi.org/10.1007/s10838-018-9439-7

Petrović A, Marković SB (2012) The cycles of revolution: how Wegener and Milanković changed the earth sciences. Acta Geogr Slov 52:259–276. https://doi.org/10.3986/AGS52200

Rundić L (2012) Centenary anniversary of the theory of continental drift by Alfred Wegener and its significance for geosciences and the human society. Bull Nat Hist Mus 5:21–23

Scotese CR (2016) PALEOMAP PaleoAtlas for GPlates and the PaleoData plotter program

Stephan T, Enkelmann E, Kroner U (2023) Analyzing the horizontal orientation of the crustal stress adjacent to plate boundaris. Sci Rep 13:15590. https://doi.org/10.1038/s41598-023-42433-2

Stern RJ (2018) The evolution of plate tectonics. Philos Trans Royal Soc A 376:20170406. https://doi.org/10.1098/rsta.2017.0406

Suárez G, García-Acosta V (2021) The first written accounts of pre-Hispanic earthquakes in the Americas. Seismol Res Lett 92:3856–3864. https://doi.org/10.1785/0220210161

Vermeesch P (2023) geostats: An Introduction to Statistics for Geoscientists

Wegener A (1920) Die Entstehung der Kontinente und Ozeane. In: Friedr. Vieweg & Sohn. Springer, Braunschweig

Wegener A (1966) The origin of continents and oceans. Dover Publications, Inc., New York

Zatman S, Gordon RG, Mutnuri K (2005) Dynamics of diffuse oceanic plate boundaries: insensitivity to rheology. Geophys J Int 162:239–248. https://doi.org/10.1111/j.1365-246X.2005.02622.x

Zilio LD, Ampuero J-P (2023) Earthquake doublet in Turkey and Syria. Commun Earth & Environ 4:71. https://doi.org/10.1038/s43247-023-00747-z

9 Land and Forest Cover Change Mapping

Ky Son, Vietnam, 2022, photo: Ł. Pawlik

> *Forests have existed for far longer than humans and have already survived a wide range of past changes in climate conditions.*
>
> (Trumbore et al. 2015)

Supplementary Information The online version contains supplementary material available at https://doi.org/10.1007/978-3-031-89673-6_9.

Land cover features and their changes are an important part of global knowledge and influence environmental management processes, political decisions, and planning under the auspices of international organizations. Mapping of land use/land cover changes is performed at several spatial and temporal resolutions, and during the past century, the process has experienced considerable steps forward, first, with the introduction of aerial photography and, second, with the development of satellites and remote sensing techniques (Senf and Seidl 2021; García-Álvarez et al. 2022; Song

© The Author(s), under exclusive license to Springer Nature Switzerland AG 2025

Ł. Pawlik, *R Applications in Earth Sciences*, Springer Textbooks in Earth Sciences, Geography and Environment,
https://doi.org/10.1007/978-3-031-89673-6_9

2023; Bonannella et al. 2024). The latter has led to the availability of large quantities of data at various spatial and spectral resolutions, but the problem remains obtaining the desired information in a precise and automatic way to support continuous change detection and final product display in a digital map form. Currently, the accuracy of machine learning and deep learning algorithms in land use and land cover change detection has reached >80% (Sefrin et al. 2021; Yuh et al. 2023). At the global scale, the resolution of satellite images is generally between 3 and 30 m (for example, Landsat, Sentinel, and PlanetScope), and ready-to-use products are available through dynamic web map services (Hansen et al. 2013). With the rapid advancement of modern technology, other types of data are now available and can be used for land cover changes, especially deforestation and afforestation. They include data produced by laser scanning from satellites, air planes, and drones (unmanned aerial vehicles, UAVs). These data are provided as point clouds that provide information on laser beam signals reflected from ground surfaces, plants, buildings, etc. Points of known positions (x-coordinates, y-coordinates, and elevations) require classification to obtain correct information on the Earth's surface and its features.

The whole procedure of land cover and land use detection and subsequent mapping requires at least the following components:

1. Technology that supports the collection of specific data (satellites, aircraft, or drones with sensors, scanners and/or cameras), access to the data through internet connections, and data storage and processing (computer hard disc memory, high-speed processors).
2. High-quality data at optimal spatial (high spatial resolution), temporal (collected every fixed number of days), and spectral resolutions (optical or radar domain), with global coverage.
3. Free and open access to the data (normally possible for noncommercial applications).
4. Algorithms that support data preprocessing, modeling, and analyses.
5. Software for script writing and control of data processing steps.
6. Mapping and map preparation in the form of thematic maps that display the final products.
7. Communication with end users through a web map service.

9.1 Description of the Data and Project Outline

We use satellite imagery from the Sentinel 2 mission launched by the European Space Agency in 2015. Notably, Sentinel provides satellite imagery in the optical domain. The Sentinel 2 mission comprises two satellites that have swath widths of 290 km and a 5-day revisit period under cloud-free conditions (https://sentinel.esa.int/web/sentinel/missions/sentinel-2). Sentinel-2 is an instrument that samples 13 spectral bands: four bands at a **10 m** spatial resolution (B2 (*490 nm*), B3 (*560 nm*), B4 (*665 nm*), and B8 (*842 nm*)); six bands at a **20 m** spatial resolution (B5 (*705 nm*), B6 (*740 nm*), B7 (*783 nm*), B8a (*865 nm*), B11 (*1610 nm*), and B12 (*2190 nm*)); and three bands at a **60 m** spatial resolution (B1 (*443 nm*), B9 (*940 nm*), and B10 (*1375 nm*)) (Table 9.1). B2, B3, and B4 are the three classical RGB bands (B2—blue, B3—green, and B4—red). B8 is a near-infrared band. B5-B8a are four narrow bands in the visible and near-infrared (VNIR) vegetation red-edge spectral domain (Clevers and Gitelson 2013). The two widest bands, the shortwave infrared (SWIR) bands, are dedicated to the detection of snow, ice, and clouds as well as the moisture levels of vegetation, such as for stress assessments (Park et al. 2017). The three 60 m bands are used for cloud screening, atmospheric correction, and cirrus detection (https://sentiwiki.copernicus.eu/web/s2-mission). For this task, we use Sentinel 2 level-2A products, which provide atmospherically corrected surface reflectance values in planar geometric form. These products are open access and free of charge. Using the Sentinel 2 images, we calculated the normalized difference vegetation index (NDVI) for the pre- and post-landslide (debris flow) images.

```
pacman::p_load("rsi", "gt", "terra", "sf",
"tmap", "tmaptools")

#Here is a table from Clevers and Gitelson
(2013) with information on spectral bands of
the Sentinel 2 images.

sentinel2 <- data.frame(sb = c("B1", "B2",
"B3", "B4", "B5", "B6", "B7", "B8", "B8a",
"B9", "B10", "B11", "B12"),
                     cw = c(443, 490, 560, 665,
                     705, 740, 783, 842, 865,
                     945, 1380, 1610, 2190),
                     bw = c(20, 65, 35, 30, 15,
                     15, 20, 115, 20, 20, 30,
                     90, 180),
```

Table 9.1 The Sentinel-2 satellite system spectral bands and their specifications (Clevers and Gitelson 2013)

Spectral band	Center Wavelength (nm)	Band width (nm)	Spatial resolution (m)
B1	443	20	60
B2	490	65	10
B3	560	35	10
B4	665	30	10
B5	705	15	20
B6	740	15	20
B7	783	20	20
B8	842	115	10
B8a	865	20	20
B9	945	20	60
B10	1380	30	60
B11	1610	90	20
B12	2190	180	20

```
                    sr = c(60, 10, 10, 10, 20,
                    20, 20, 10, 20, 60, 60,
                    20, 20))
```

```
#The column names can be changed in this way.
```

```
colnames(sentinel2) <- c("Spectral band",
"Center Wavelength (nm)", "Band width (nm)",
"Spatial resolution (m)")
```

```
#Now, we will modify and save the table using
the gt package.
```

```
library(gt)
```

```
sentinel2a <- sentinel2 %>%
  gt() %>%
  cols_width(everything() ~ px(100)) %>%
  cols_align(align = "center") %>%
  data_color(columns = "Spatial resolution
(m)",
             palette = "Spectral")
sentinel2a
```

```
#We can save the table.
```

```
#gtsave(sentinel2a, "sentinel2_table.png")
```

9.1.1 R in Practice: Changes in Land Cover Caused by Landslides in Northern Vietnam

Several R packages support the downloading and processing of satellite images, such as the **sits** package (Simoes et al. 2021), the **satellite** package (Nauss et al. 2024), the **MODISTools** package (Hufkens 2023), and the **rgee** package (Aybar 2023). However, to introduce satellite image applications in the context of land cover change, we use the **rsi** package (Mahoney 2024). One advantage of this package is that it simplifies the process of obtaining spatial data from the **SpatioTemporal Asset Catalogs** (STAC) (Rydzik 2024). The STAC is a standard and a community of collaborators working to enable increased access to information about planet Earth. The focus is to provide easy access to global data (https://stacspec.org/en/). The STAC specification provides a common structure for describing and cataloging spatiotemporal assets (i.e., any file that represents information about the Earth's surface captured in a certain space and time). The **rsi** package supports data download and provides a wealth of information about spectral indices from the **Awesome Spectral Indices** repository, with nearly 250 indices (https://awesome-ee-spectral-indices.readthedocs.io/en/latest/#). We use the NDVI to evaluate changes caused by heavy rainfall that triggered landslides and debris flows in one of the regions of northern Vietnam. The NDVI expresses the health condition of vegetation (Rouse Jr. et al. 1974) and is calculated as follows: NDVI = NIR—RED/NIR + RED, where NIR refers to the near-infrared band and RED refers to the red band in the visible region (Labib et al. 2020). Because we use Sentinel-2 images, the translation of this algorithm into specific bands is B8—B4/B8 + B4. The NDVI values are between −1 and 1, and high positive values indicate a proportion of green vegetation coverage. Studies of global vegetation changes at northern latitudes (over 50°N) have shown that between 1982 and 2015, forests had the highest NDVI values, ranging from 0.79 to 0.84, whereas croplands had values ranging from 0.69 to 0.75 and shrublands had values ranging from 0.49 to 0.73 (Shou-Ye et al. 2021). The NDVI in the forest and tundra areas displayed increasing trends, but this trend was statistically significant only for the tundra biome (p value <0.05).

Northern Vietnam is highly susceptible to landslide movements and formation because of several overlapping triggering factors, such as high levels of precipitation during the monsoon season and tropical cyclones, steep slopes, and deeply weathered bedrock. The North Vietnam territory can receive more than 3000 mm of rainfall annually, and up to 15% of this rainfall is associated with tropical cyclones (Pham-Thanh et al. 2020). During landslide events, vegetation is normally removed via the downslope movement of the soil and weathered rock masses. Vegetation can also be significantly damaged or covered by deposits (so-called colluvium) transported during mass movement. For this reason, the growth conditions and distribution frequently dramatically change in areas experiencing widespread and reoccurring landslide events. Figure 9.1 shows examples from one study site in Hà Giang Province, northern Vietnam, where intense rainfall triggered landslides on 16 October 2022. Several landslides formed simultaneously, transporting large volumes of soil and weathered material along with vegetation cover.

```
library(rsi)
library(terra)

#Information from Awesome Spectral Indeces
library (https://awesome-ee-spectral-indices.
readthedocs.io/en/latest/)


#The website can be opened in a new default
browser card using the browseURL() function.
#browseURL("https://awesome-ee-spectral-
indices.readthedocs.io/en/latest/")

ASI <- spectral_indices(download_indices =
TRUE)
ASI

library(sf)


#The definition of the spatial range of the area
of interest.
#The geographical coordinates are expressed in
meters in the UTM (Universal Transverse
Mercator) coordinate system.

points_df <- data.frame(long = c(478834.2,
478753.9, 474499.8, 474499.8, 478834.2), lat =
c(2493293, 2489802, 2489802, 2493293, 2493293))


#Points are now converted to a polygon.
#Please see this website for the explanation of
the EPSG (European Petroleum Survey Group) code
#https://spatialreference.org/ref/epsg/32648/

polygon <- points_df %>%
  st_as_sf(coords = c("long", "lat"), crs =
32648) %>%
  st_combine() %>%
  st_cast("POLYGON")

st_coordinates(polygon)


#Satellite images for the period before the
landslide formation.


xuan_minh_before_landslide <- get_sentinel2_
imagery(
  polygon,
```

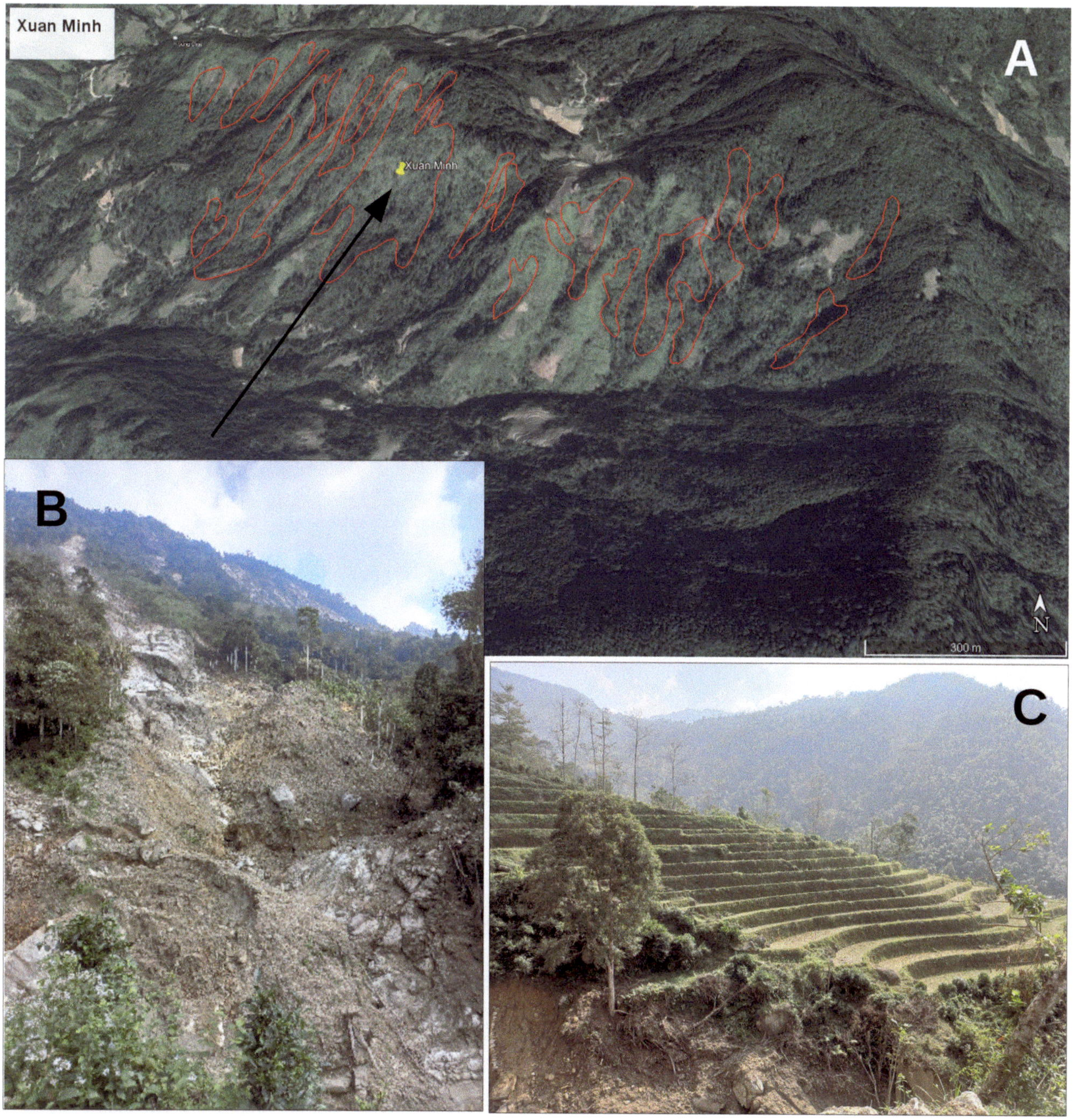

Fig. 9.1 Landslide formation and effects documented in Hà Giang Province. A: Landslides mapped during the fieldwork and displayed in Google Earth satellite imagery. B: A landslide scar and transport zone with completely removed vegetation. C: Rice terraces partly damaged by the landslide. Photographs B and C: Ł. Pawlik

```
  start_date = "2022-09-03",
  end_date = "2022-09-10",
  output_filename = tempfile(fileext = ".tif"),
  composite_function = NULL
)

#We got one image for this period. It can be
selected and plotted as a raster layer. Before
the image is transformed into a SpatRaster
object using the rast function from the terra
package.


#05-09-2022
#png("xuan_minh05092022.png", width = 1000,
height = 600)
plot(rast(xuan_minh_before_landslide[1]))#plot
all bands
```

```
#dev.off()

#Plot selected band.
#plot(rast(xuan_minh_before_landslide[1])$A)

#Labeling of the Sentinel 2 bands according to
the STAC server.
sentinel2_band_mapping$planetary_computer_v1

#Images for the period after the landslide
formation.


xuan_minh_after_landslide <- get_sentinel2_
imagery(
  polygon,
  start_date = "2022-11-08",
  end_date = "2022-11-30",
  output_filename = tempfile(fileext = ".tif"),
  composite_function = NULL
)

#You can inspect the object by simply typing
its name.
xuan_minh_after_landslide

#Select an image from 29 November 2022.
#plot(rast(xuan_minh_after_landslide[1]))

#We select the NDVI algorithm from the ASI
catalog.
NDVI <- ASI[ASI$short_name == "NDVI",]

#NDVI for 05-09-2022.
xuan_minh_before_landslide_ndvi <- calculate_
indices(
  xuan_minh_before_landslide[1],
  NDVI,
  output_filename = tempfile(fileext = ".tif")
)

#NDVI for 29-11-2022.
xuan_minh_after_landslide_ndvi <- calculate_
indices(
  xuan_minh_after_landslide[1],
  NDVI,
  output_filename = tempfile(fileext = ".tif")
)

#Plot the images with the tmap package.


library(tmap)
library(tmaptools)

#Define map color scale.
#tmaptools::palette_explorer()

#This function starts an interactive tool that
helps to select the appropriate color palette
from the Color Brewer and viridis palettes.
#More about colors: https://www.color-
blindness.com/types-of-color-blindness/

#png("color_scale.png", width = 800, height =
200)
tmaptools::get_brewer_pal("RdYlGn", n = 10)
#dev.off()

#NDVI map 1.
ndvi05092022
<- tm_shape(rast(xuan_minh_before_landslide_
ndvi))+
  tm_raster(col.scale = tm_scale_
  continuous(values = c(tmaptools::get_brewer_
  pal("RdYlGn", n = 10, plot = FALSE))))+
  tm_title("NDVI - 05-09-2022")+
  tm_grid(labels.rot = c(90,0))+
  tm_compass(size = 3, cardinal.directions =
  c("N", "E", "S", "W"),
             type = "8star", position =
             c(0.15,0.35))+
  tm_scalebar(position = c(0.1,0.1))+
  tm_layout(legend.show = TRUE,
            legend.position = c("right",
            "top"))

#NDVI map 2.
ndvi29112022
<- tm_shape(rast(xuan_minh_after_landslide_
ndvi))+
  tm_raster(col.scale = tm_scale_
  continuous(values = c(tmaptools::get_brewer_
  pal("RdYlGn", n = 10, plot = FALSE))))+
  tm_title(("NDVI - 29-11-2022")+
  tm_grid(labels.rot = c(90,0))+
  tm_compass(size = 3, cardinal.directions =
  c("N", "E", "S", "W"),
             type = "8star", position =
             c(0.15,0.35))+
  tm_scalebar(position = c(0.1,0.1))+
  tm_layout(legend.show = TRUE,
            legend.position = c("right",
            "top"))

#Plot only the legend and add it to maps.
ndvi_legend <- tm_shape(rast(xuan_minh_after_
landslide_ndvi))+
```

```
  tm_raster(col.scale = tm_scale_
  continuous(values = c(tmaptools::get_brewer_
  pal("RdYlGn", n = 10, plot = FALSE))))+
 tm_layout(legend.only = TRUE)

#Arrange the maps and the legend in one figure.
ndvi_combined <- tmap_arrange(ndvi05092022,
ndvi29112022,
                              ncol = 2, nrow =
                              1, asp = NA)
#Save the figure.
#tmap_save(ndvi_combined, "ndvi_combined1.png",
dpi = 800, height = 10, width = 22, units =
"cm", asp = NA)
```

With knowledge of the exact time of the landslide event and the approximate limit of the landslide-affected area, we can narrow the spatiotemporal search window when the ***get_sentinel2_imagery()*** function is used and significantly accelerates the analysis. Notably, the Sentinel-2 imagery is satellite imagery in the optical domain. This means that the availability of high-quality images is limited to daytime and cloudless weather conditions. The latter poses a significant problem in Vietnam during the monsoon season from May to September. However, high-quality imagery was available on certain days more than 1 month before and after the event (Fig. 9.2). The NDVI was selected from the built-in list of the available spectral indices and applied using the ***calculate_indices()*** function. The difference between the two images is striking, showing the greatest changes on the hillslopes affected by landslides and debris flows and along the river channels where erosion and deposition took place (Fig. 9.3). To best display the results, testing various color scales may be a good option (Fig. 9.4).

9.1.2 R in Practice: Deforestation of the Amazonia Basin

In our second example, we select information on forest change as the subject of the analysis. The information is updated annually on the Global Forest Change (GFC) web service (https://glad.earthengine.app/view/global-forest-change#bl=off;old=off;dl=1;lon=20;lat=10;zoom=3;). The dataset was created by analyzing Landsat images at a 30-meter spatial resolution (Hansen et al. 2013). From this global dataset, we selected one tile covering a part of the Amazon River basin and calculated the area of forest damage each year (Fig. 9.5). Forest cover loss is defined as a stand-replacement disturbance or a change from a forest state to a nonforest state each year during the period from 2000–2023. In the GFC mapping project, when an area with >50% crown cover was degraded to ~0% crown cover, it was classified into the forest loss class. With the specific link from the GFC web service, we saved the tile as a temporary file, plotted it and calculated the damage area for each year (Fig. 9.6). Deforestation in the Brazilian Amazon region is well known. On the basis of data from 2001-2018, it was estimated that 5.5% (0.36×10^6 km^2) of the Amazon forest is experiencing some form of degradation (Lapola et al. 2023). The primary drivers of disturbances include extreme drought, fire, timber extraction, and edge effects linked to road and settlement development. These disturbances have also been especially intense in the southern and eastern parts of the region, where they threaten Indigenous territories (Silva-Junior et al. 2023). According to the GFC data, the greatest forest loss occurred in 2016 (Fig. 9.7).

```
#For the next task, we can download the data
from the web page of the Global Forest Change:
#shell.exec("https://storage.googleapis.com/
earthenginepartners-hansen/GFC-2023-v1.11/
download.html")

#We will save the file in the session memory.
destfile1
<- tempfile(pattern = "Hansen_GFC-2022-v1.10_
lossyear_00N_070W",
 tmpdir = tempdir(),
 fileext = ".tif")

#Define the URL.
url_forest <- "https://storage.googleapis.com/
earthenginepartners-hansen/GFC-2023-v1.11/
Hansen_GFC-2023-v1.11_lossyear_00N_070W.tif"

#Download the data.
download.file(url_forest, destfile1, quiet =
TRUE, method = "curl")

#We read the file as a raster layer.
amazon <- rast(destfile1)

#Configure the bounding box, which will show the
exact position of the imported tile.


amazon_bbox <- amazon %>%
 sf::st_bbox() %>%
 st_as_sfc() %>%
 st_cast("POLYGON")

library(rnaturalearth)
library(tmap)

coastlines <- ne_coastline()
```

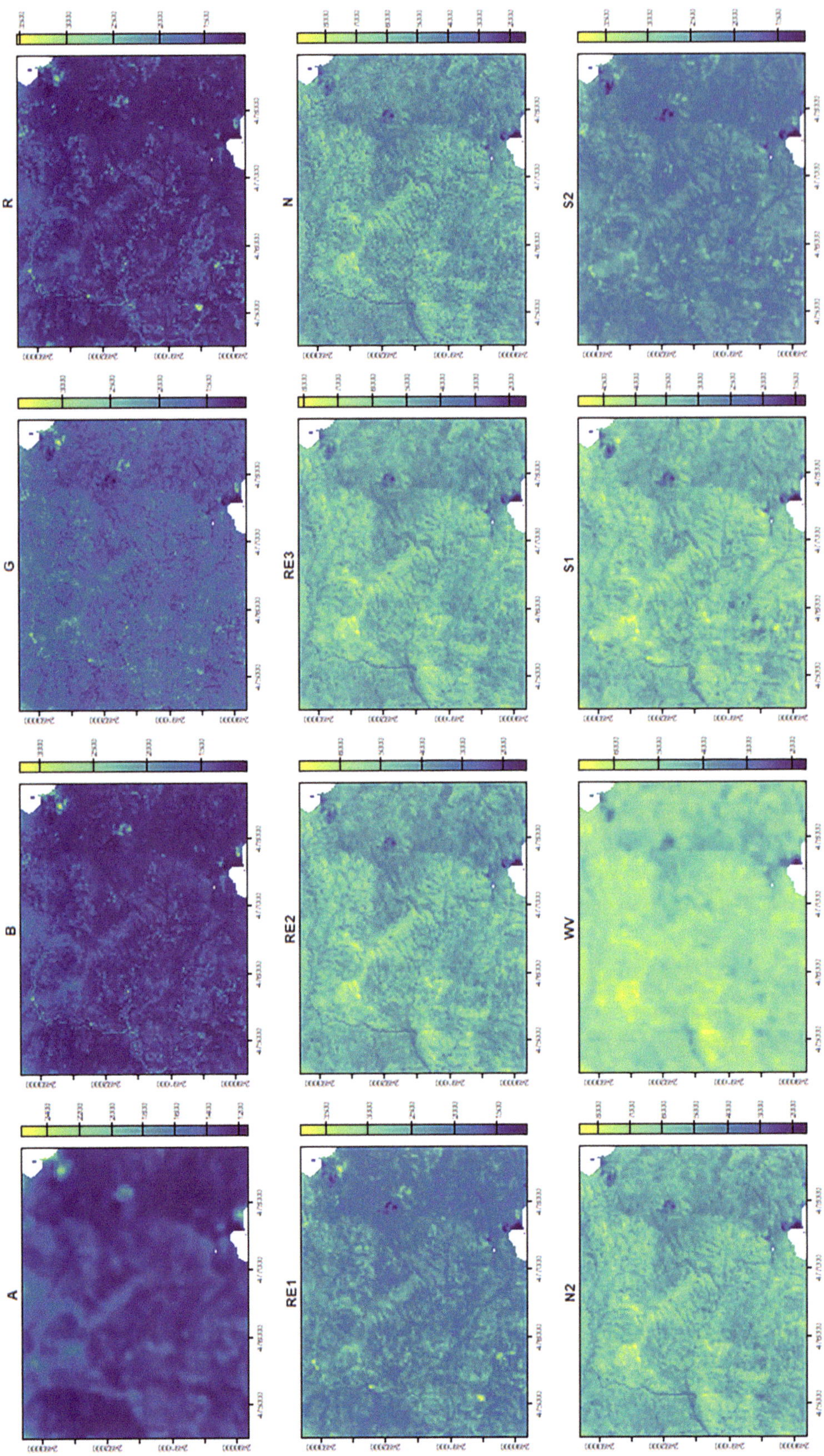

Fig. 9.2 Sentinel-2 bands with labels according to the STAC server: A—B10, B—B02, G—B03, R—B04, RE1—B05, RE2—B06, RE3—B07, N—B08, N2—B08A, WV—B09, S1—B11, and S2—B12. The list of band labels can be obtained using sentinel2_band_mapping$planetary_computer_v1

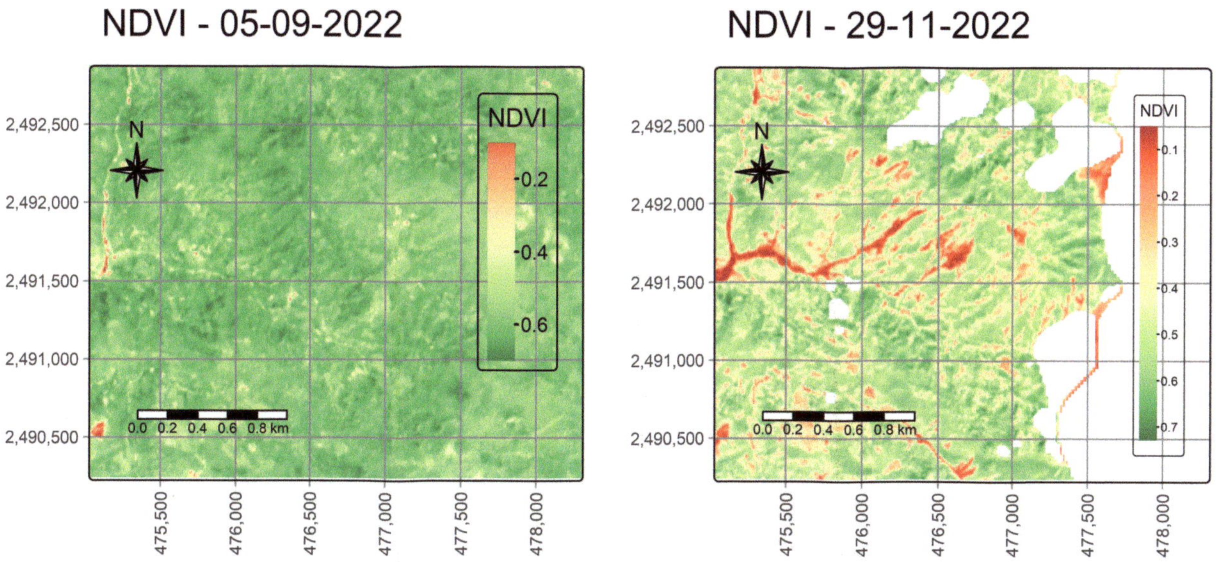

Fig. 9.3 The NDVI calculated for the study area located in northern Vietnam. The cloudless image obtained for the period before the landslide event is from 5 September 2022. The image for the period after the landslide formation event is from 29 November 2022. The maps were prepared with the **tmap** package

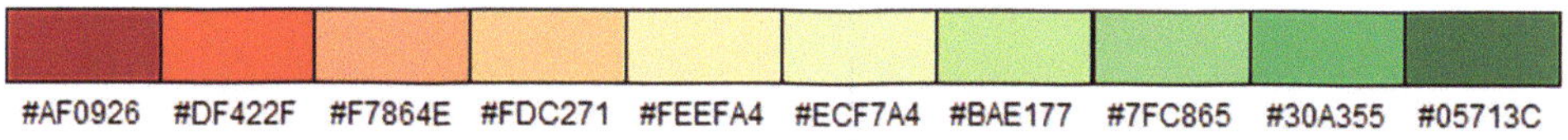

Fig. 9.4 An example of the color scale defined with ***get_brewer_pal()*** from the **tmaptools** package (Tennekes 2021). Before using the function, available color palettes can be inspected using the ***palette_explorer()*** function, which opens the Shiny application

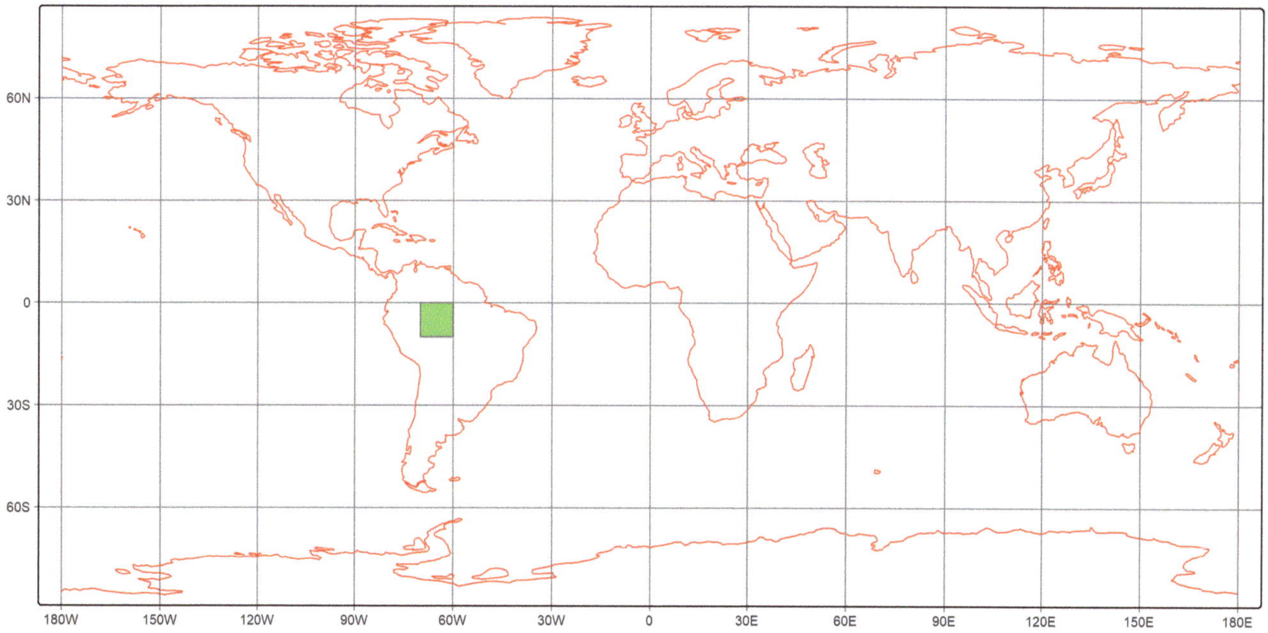

Fig. 9.5 The area of interest plotted on the world map. The coastlines are from the **rnaturalearth** package (Massicotte and South 2023), and the bounding box of the study area (green square) was obtained using the ***sf::st_bbox()*** function

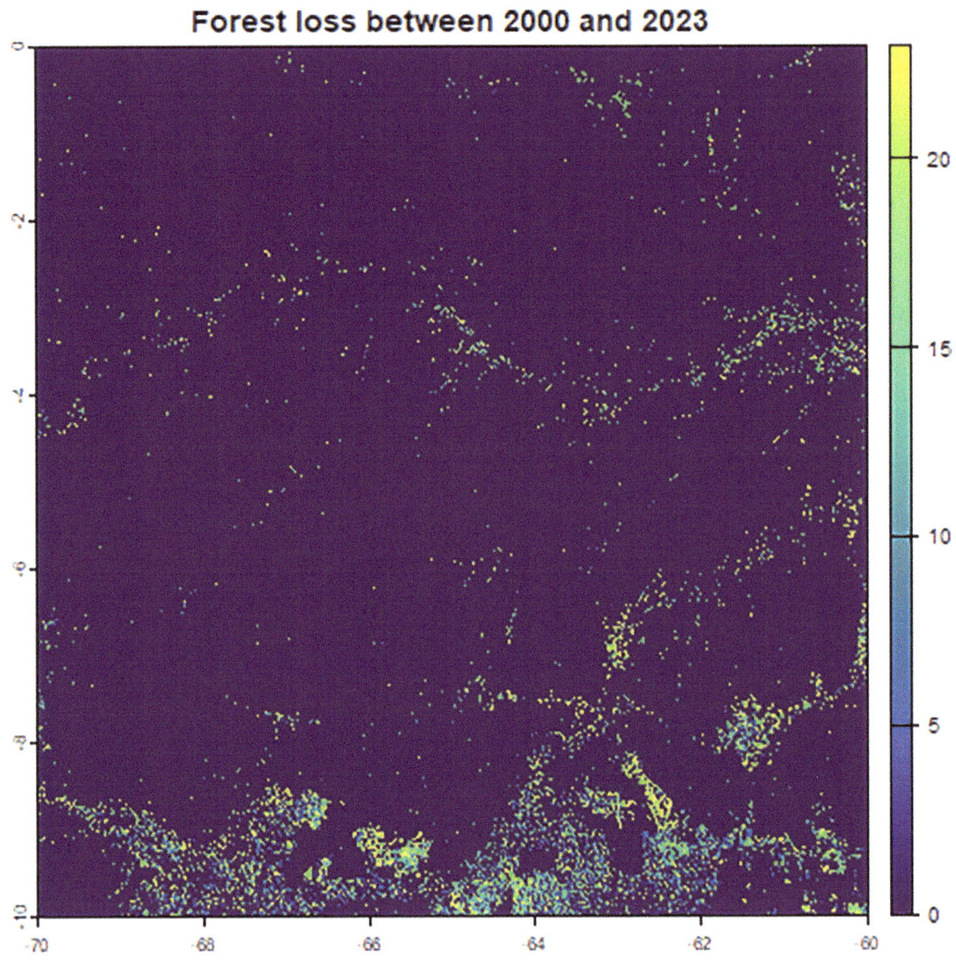

Fig. 9.6 Map displaying forest loss for each year between 2001 and 2023. The changes were encoded as either 0 (no loss) or a value in the range of 1–23, representing losses detected primarily from 2001–2023, respectively

Fig. 9.7 Deforestation of the Amazon forest is based on the results of Hansen et al. (2013). The data are annually updated from the Global Forest Change web service (https://glad.earthengine.app/view/global-forest-change#bl=off;old=off;dl=1;lon=20;lat=10;zoom=3;)

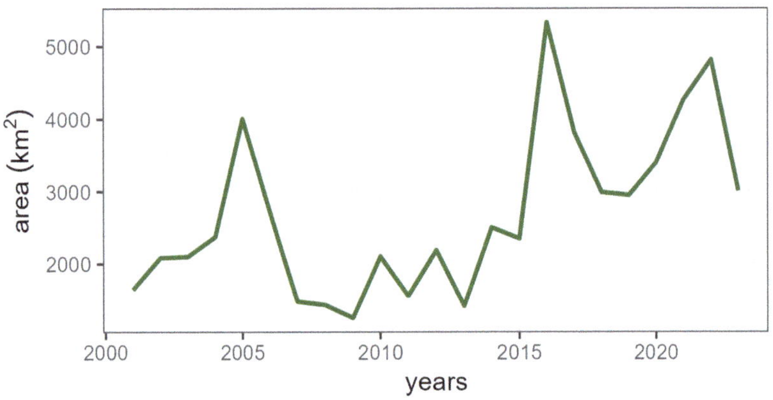


```
hansen_tile <- tm_shape(coastlines$geometry)+
 tm_graticules()+
 tm_lines(col = "red")+
 tm_shape(amazon_bbox)+
 tm_polygons(fill = "green")

#Plot the map.
hansen_tile

#Save the map.
#tmap_save(hansen_tile, "hansen_data_tile.png")
```

```
#Plotting.
#png("hansen_tile_00N_070W.png")
plot(amazon, main = "Forest loss between 2000
and 2023")
#dev.off()

#When was the most significant loss?
#We need to calculate area by category; each
category is the year when forest loss was
recorded.

#The expanse() function computes the area
covered by raster cells that are not NA. The
byValue argument was set to TRUE, so each
year's area was calculated separately.


area1 <- terra::expanse(amazon, unit = "km",
byValue = TRUE)


area2 <- data.frame(year = 2001:2023, area =
area1[-1,3])


library(ggplot2)


ggplot()+
  geom_line(area2, mapping = aes(x=year,
  y=area), size=0.9, color="darkgreen")+
  xlab("years")+
  ylab(expression(area ~ (km^2)))+
  theme_bw()

#Save the plot.
#ggsave("amazonia_area_changes.png", width =
12, height = 6, units = "cm")
```

References

Aybar C (2023) rgee: R bindings for calling the "earth engine" API

Bonannella C, Parente L, de Bruin S, Herold M (2024) Multi-decadal trend analysis and forest disturbance assessment of European tree species: concerning signs of a subtle shift. For Ecol Manag 554:121652. https://doi.org/10.1016/j.foreco.2023.121652

Clevers JGPW, Gitelson AA (2013) Remote estimation of crop and grass chlorophyll and nitrogen content using red-edge bands on Sentinel-2 and -3. Int J Appl Earth Obs Geoinf 23:344–351. https://doi.org/10.1016/j.jag.2012.10.008

García-Álvarez D, Olmedo MTC, Mas J-F, Paegelow M (2022) Land use cover mapping, modelling and validation. A background. In: Land use cover datasets and validation tools. Validation practices with QGIS. Springer, pp 21–34

Hansen MC, Potapov PV, Moore R et al (2013) High resolution global maps of 21st-century forest cover change. Science 342:850–853. https://doi.org/10.1126/science.1244693

Hufkens K (2023) The MODISTools package: an interface to the MODIS land products subsets web services

Labib SM, Lindley S, Huck JJ (2020) Scale effects in remotely sensed greenspace metrics and how to mitigate them for environmental health exposure assessment. Comput Environ Urban Syst 82:101501. https://doi.org/10.1016/j.compenvurbsys.2020.101501

Lapola DM, Pinho P, Barlow J et al (2023) The drivers and impacts of Amazon forest degradation. Science 379:eabp8622. https://doi.org/10.1126/science.abp8622

Mahoney M (2024) rsi: efficiently retrieve and process satellite imagery

Massicotte P, South A (2023) rnaturalearth: world map data from natural earth

Nauss T, Meyer H, Appelhans T, Detsch F (2024) Satellite: handling and manipulating remote sensing data

Park H, Choi J, Park N, Choi S (2017) Sharpening the VNIR and SWIR bands of sentinel-2A imagery through modified selected and synthesized band schemes. Remote Sensing 9:1080. https://doi.org/10.3390/rs9101080

Pham-Thanh H, Ngo-Duc T, Matsumoto J et al (2020) Rainfall trends in Vietnam and their associations with tropical cyclones during 1979-2019. SOLA 16:169–174. https://doi.org/10.2151/sola.2020-029

Rouse JW Jr, Haas RH, Schell JA, Deering DW (1974) Monitoring vegetation systems in the great plains with ERTS. NASA, pp 309–317

Rydzik M (2024) An overview of the Rsi R package for retrieving Sattelite imagery and calculating spectral indices. In: geocompx. https://geocompx.org/post/2024/rsi-bp1/

Sefrin O, Riese FM, Keller S (2021) Deep learning for land cover change detection. Remote Sens 13:78. https://doi.org/10.3390/rs13010078

Senf C, Seidl R (2021) Mapping the forest disturbance regimes of Europe. Nat Sustain 4:63–70. https://doi.org/10.1038/s41893-020-00609-y

Shou-Ye X, Hai-Yan X, Cui-Cui M et al (2021) Changes in different land cover areas and NDVI values in northern latitudes from 1982 to 2015. Adv Clim Chang Res 12:456–465. https://doi.org/10.1016/j.accre.2021.04.003

Silva-Junior CHL, Silva FB, Arisi BM et al (2023) Brazilian Amazon indigenous territories under deforestation pressure. Sci Rep 13:5851. https://doi.org/10.1038/s41598-023-32746-7

Simoes R, Camara G, Souza F et al (2021) Satellite image time series analysis for big earth observation data. Remote Sens 13:2428. https://doi.org/10.3390/rs13132428

Song X-P (2023) The future of global land change monitoring. Int J Digit Earth 16:2279–2300. https://doi.org/10.1080/17538947.2023.2224586

Tennekes M (2021) tmaptools: thematic map tools

Trumbore S, Brando P, Hartmann H (2015) Forest health and global change. Science 349:814–818. https://doi.org/10.1126/science.aac6759

Yuh YG, Tracz W, Matthews HD, Turner SE (2023) Application of machine learning approaches for land cover monitoring in northern Cameroon. Ecol Indic 74:1011955. https://doi.org/10.1016/j.ecoinf.2022.101955

10 Extreme Climate Event Modeling and Prediction with Machine Learning Methods

Karkonosze National Park, Sudety Mountains, Poland, 2024, photo: Ł. Pawlik

Supplementary Information The online version contains supplementary material available at https://doi.org/10.1007/978-3-031-89673-6_10.

© The Author(s), under exclusive license to Springer Nature Switzerland AG 2025

Ł. Pawlik, *R Applications in Earth Sciences*, Springer Textbooks in Earth Sciences, Geography and Environment,
https://doi.org/10.1007/978-3-031-89673-6_10

Natural disturbances have complex effects on forest ecosystems.

(Machado Nunes Romeiro et al. 2022)

Extreme events are those that occur relatively rarely and cause large or key changes to a system (Stewart et al. 2022). These events are associated with an exceptional situation and are defined by a threshold value that, when crossed, results in the system's inability to accommodate the effects of the change. Consequently, the system may become unstable. The knowledge of threshold values or levels that define the extremity of an event is based on long-term observations (climate systems), fossil evidence (glaciations, megafloods, and tsunamis), or projections of the current data into the past or future (global climate warming and cooling). For example, from a statistical viewpoint, an extreme climatic event is defined as "*an episode or occurrence in which a statistically rare or unusual climatic period alters ecosystem structure and/or function well outside the bounds of what is considered typical or normal variability*" (Smith 2011). Any extreme event can be defined in terms of its rarity, which is calculated as a percentile value (for instance, 0.95 or 0.99) that separates the values above and below (Beniston and Stephenson 2004). This value will be, of course, different for different locations. Extreme events associated with natural hazards are also defined by the level of damage caused to man-made infrastructure and the number of fatalities (Chaudhary and Piracha 2021). These events are also referred to as catastrophes (Svensen 2009; Donatti et al. 2024). Different branches of science may have unique definitions of an extreme event or extremity. In geosciences, extreme events are seen as the effects of many complex, nonlinear systems in which various interdependent components and their interactions lead to disruptions in system balance and irregular behaviors (Surjalal Sharma et al. 2012). Such events are commonly associated with hazards and can have a significant negative impact on humans, infrastructure, or the environment (McPhillips et al. 2018). In ecology, a disturbance is related to the sudden partial or complete removal of biomass, which is a driver of change. An extreme event that leads to ecosystem changes, often stimulating a system to further develop, such as after a wildfire or windstorm, or causing its partial or complete destruction (e.g., deforestation, land cover change conversion to crop fields, or continental glaciation).

In this chapter, we focus on European windstorms and their effects (Fig. 10.1). Extreme wind speed is one of the primary factors leading to severe forest damage in European forests (Brönnimann et al. 2012; Seidl et al. 2017). Extreme wind is also a potential threat to human lives (Schwierz et al. 2010; Badoux et al. 2016; Dawkins et al. 2016) and can result in landscape change and soil disturbances (Lindtroth et al. 2009; Godziek and Pawlik 2023). It was assumed that between 1950 and 2010, approximately 51% of all damage in European forests was caused by strong winds, and since the beginning of the twentieth century, a growing trend in forest damage has been observed (Schelhaas et al. 2003). It was also estimated that annual losses for the EU and UK were as high as 5 billion €, or approximately 0.04% of the total GDP (Spinoni et al. 2020). In Germany, 53% of economic losses related to natural hazards are caused by winter storms (Donat et al. 2011). Between 1984 and 2016, the canopy mortality (due to various factors) of temperate forests in Europe increased by +2.4% $yr.^{-1}$ on average (Senf et al. 2018). This steady increase resulted in the doubling of the forest area affected by mortality since 1984. Many simulations have shown a marked increase in climate severity indices, such as extreme wind conditions (Ruosteenoja et al. 2019).

10.1 Description of the Data and Project Outline

We first work on an example dataset containing extratropical windstorm trajectories, which are paths of deep cyclones that typically form and develop over the North Atlantic during the winter season (DJF). They hit European territory, causing extensive damage to infrastructure and forests. The dataset displayed in Fig. 10.2 is part of the **esd** package (Benestad et al. 2024). The storm paths were obtained using a calculus-based cyclone identification (CCI) method (Neu et al. 2013). These extreme climate events are frequently the cause of forest damage in Europe. One example of such damage is the main subject of modeling in this chapter.

```
pacman::p_load("esd")
remotes::install_github("metno/esd")

library(esd)


#Extra-tropical windstorm trajectories.
data("imilast.M03")

#png("trajectories.png", width = 1000, height =
600)
par(mfrow = c(1,2))#two maps in one row

#Windstorm paths.
map(imilast.M03, it='djf', projection =
'sphere', new=FALSE, col="red", latR = 55, lonR
= -20)

#Windstorm path density.
map(imilast.M03, id='djf', type = "density",
colbar=list(pal="cold", rev=TRUE),
```

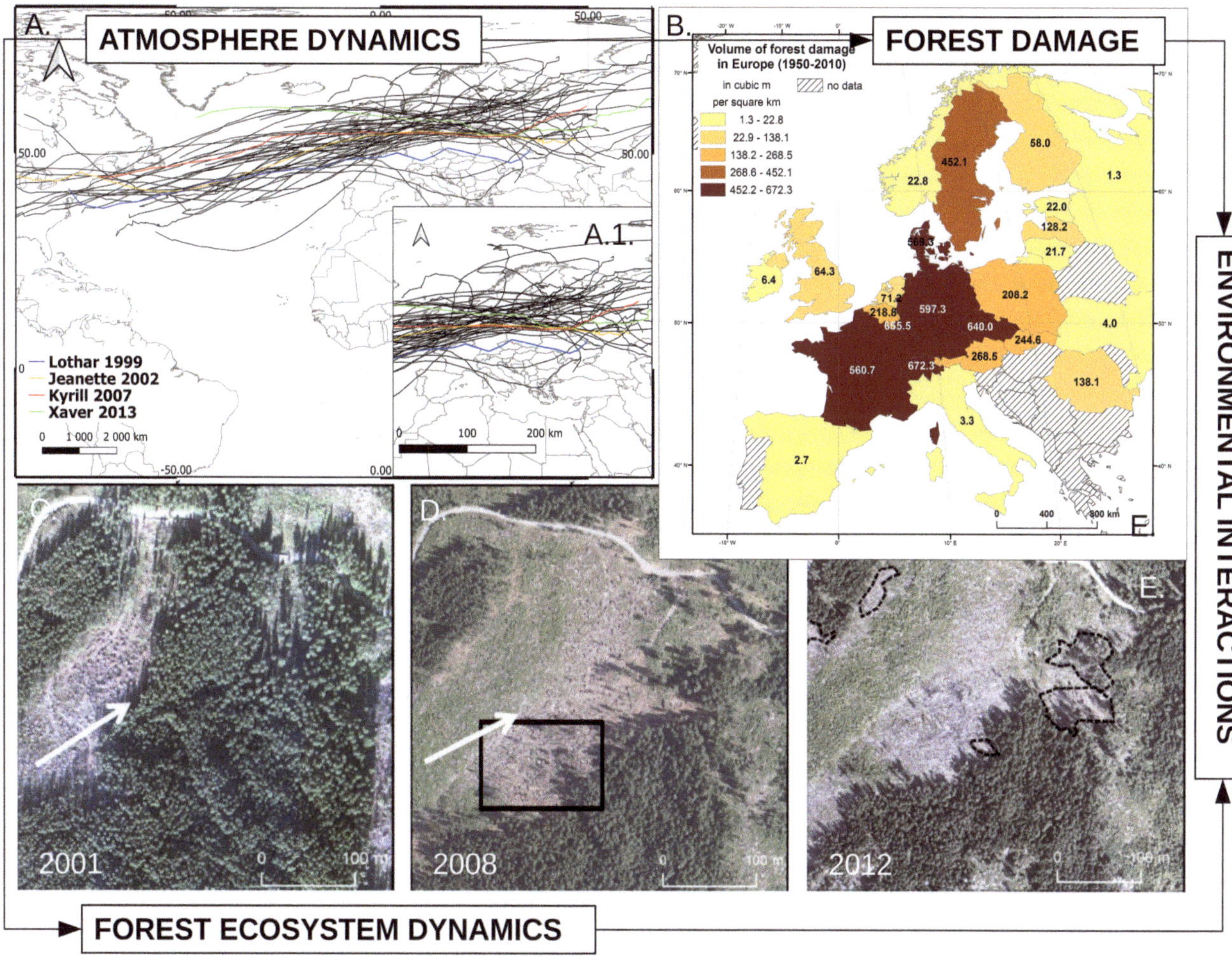

Fig. 10.1 Extreme European windstorms and their outcomes. (**a**) Atmospheric dynamics related to extratropical winter storms over the North Atlantic and Europe (A.1). The figure is based on data from the Extreme Wind Storms Catalog (http://www.europeanwindstorms.org/) (Roberts et al. 2014). The colored lines highlight four extreme winter windstorms: Lothars 26-12-1999, Jeanette 27-10-2002, Kyrill 18-01-2007, and Xaver 5-12-2013. (**b**) Reported forest damage in Europe between 1950 and 2010. Map B is based on Gardiner et al.'s (Gardiner et al. 2010, 2013) data. (**c**) Forest damage caused by strong winds in Karkonosze National Park, SW Poland, in 2001 (**c**), after the Kyrill windstorm on 18–19 January 2007 (**d** image from 2008), and in 2012 (**e**). The white arrows indicate the approximate wind flow direction during the Kyrill windstorm (**c** and **d**). The black rectangle indicates a research plot where field measurements were conducted from 2011–2012 and reported by Pawlik et al. (2016)

```
xlim=c(-90,60), ylim=c(30,90), dx=4, dy=2,
new=FALSE)
#dev.off()
```

Datasets with information on forest damage due to strong winds are limited, and they normally need to be transformed into a more appropriate format. For example, for classification modeling, two classes of response variables are typically needed. The response variable is binary. In our case study, the two classes indicate places with and without forest damage. One database in which records of forest damage across Europe are compiled is the FORWIND dataset (Forzieri et al. 2020). It includes almost 90,000 records of areas disturbed by strong winds. Although it contains only a fraction of known historical disturbances, it can be used for modeling damage caused by strong winds in forest ecosystems in specific areas (Pawlik et al. 2022). In this section, we build a random forest (RF) classification model using a binary response variable (having only two classes, positive and negative) and a set of predictors. We test the predictive power of the best-performing model by applying it to a test dataset, which is excluded from the training procedure. Then, the importance of various predictors to the prediction results is evaluated.

Here, we analyzed one instance of a catastrophic winter wind storm (extratropical cyclone) that hit a maritime pine forest in the SW part of France on 24 January 2009 (Fig. 10.3). The damage caused by this catastrophic windstorm was estimated to have resulted in 42 Mm^3 of tree damage and, unfortunately, 26 fatalities. This specific windstorm was named Klaus. The naming convention is an interesting issue itself. Severe windstorms are named by climate agencies, such as the MetOffice in the United Kingdom, to avoid

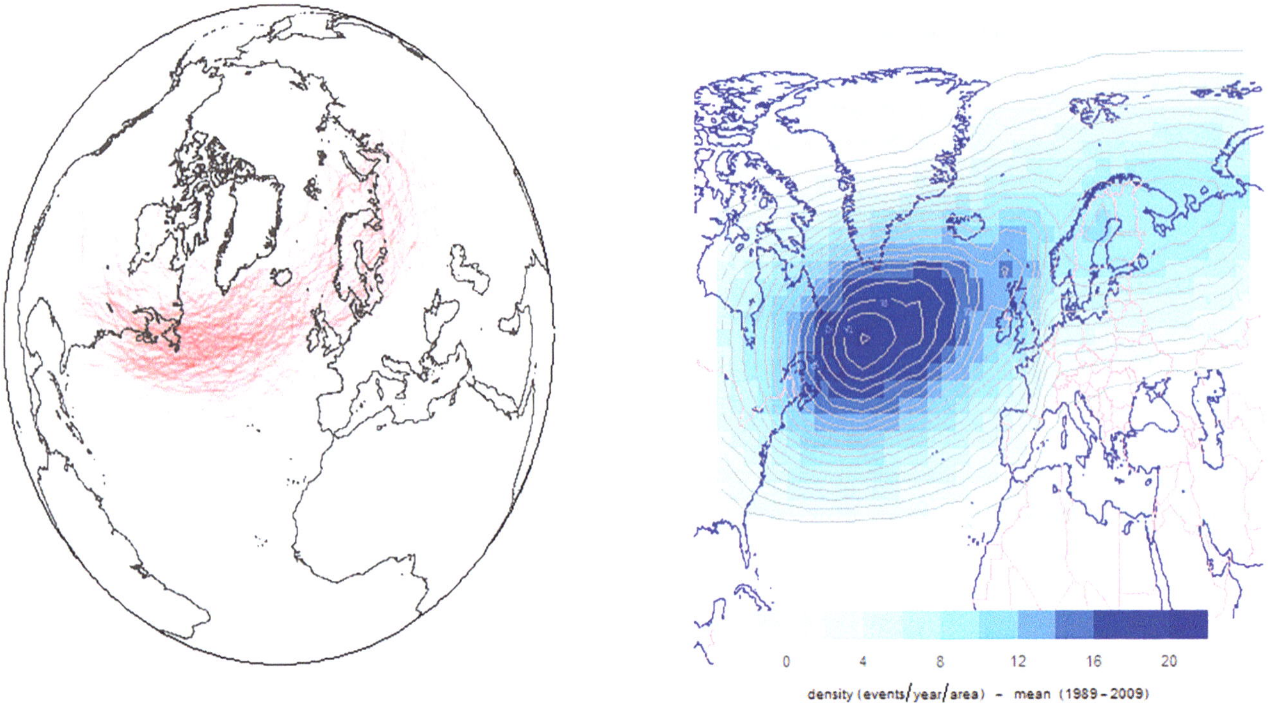

Fig. 10.2 Extratropical windstorm trajectories during the winter season (DJF) and their density per 1000 km^2 for the 1989–2009 period. The dataset is part of the **esd** package

confusion, provide a consistent message, and aid in the communication of information regarding approaching severe weather through media partners and government agencies (https://www.metoffice.gov.uk/weather/warnings-and-advice/uk-storm-centre/index). Since 1954, low- and high-pressure systems over Europe have been named by the Free University of Berlin (Charlton-Perez et al. 2019).

The dataset (in shapefile (SHP) format) we used contains information on the **response variable**: the magnitude of damage expressed as a percentage and the locations of damage (as polygons). However, to build a **predictive model**, a set of **independent variables** is needed. These variables are related to the magnitude and spatial variability of damage caused by the Klaus windstorm. The **potential predictors** were evaluated during the model training procedure, and only the most important predictors were included in the final model. For this project, the following potential predictors were selected: elevation in m above sea level (a.s.l.), slope angle (in radians), terrain roughness, profile curvature, platform curvature, wind speed (in m s^{-1}), topographic wetness index (TWI), wind exposure index (WEI), distance from the coastline (in km), distance from the windstorm track (in km), soil clay content, leaf area index (LAI), and normalized difference vegetation index (NDVI, values from −1 to 1). The level of damage is related primarily to the wind speed and gustiness, but these data are difficult to obtain, especially at high spatial resolutions (meters) and over hourly time series frequencies (hourly rather than daily values). Therefore, in the spatial prediction approach, other potential predictors are used to help explain why damage occurred in some places but not in others. Notably, the topographical features of landforms, individual tree or forest stand characteristics, soil features, and hillslope hydrology are important to consider. For example, the LAI and NDVI provide information about forest conditions before and after a windstorm. Historical observations have shown that forest damage caused by strong winds often occurs at relatively high elevations, on steep slopes, and in areas with wet soils. However, this information can be site specific and needs to be evaluated at any given location.

For the modeling procedure, we use the RF technique, which can be applied to solve both classification and regression problems (Breiman 2001). An RF model is a supervised method that requires historical observations with labeled response variables if classification is the main aim of modeling. In this method, multiple independent decision trees (individual classifiers) are established through bagging (creating multiple subsets of data). In the standard approach, all trees are treated equally, but in some cases, their responses are weighted on the basis of their individual performance metrics and averaged (aggregated) (Probst and Boulesteix 2017; Lesmeister and Chinnamgari 2019). This approach helps increase model stability and accuracy. The RF method requires hyperparameter tuning, and parameters must be set before running the algorithm (Probst et al. 2019a, b). Hyperparameter tuning (optimization) is the process of finding the optimal settings of an algorithm for a specific dataset. This approach aids in optimizing the structure of each indi-

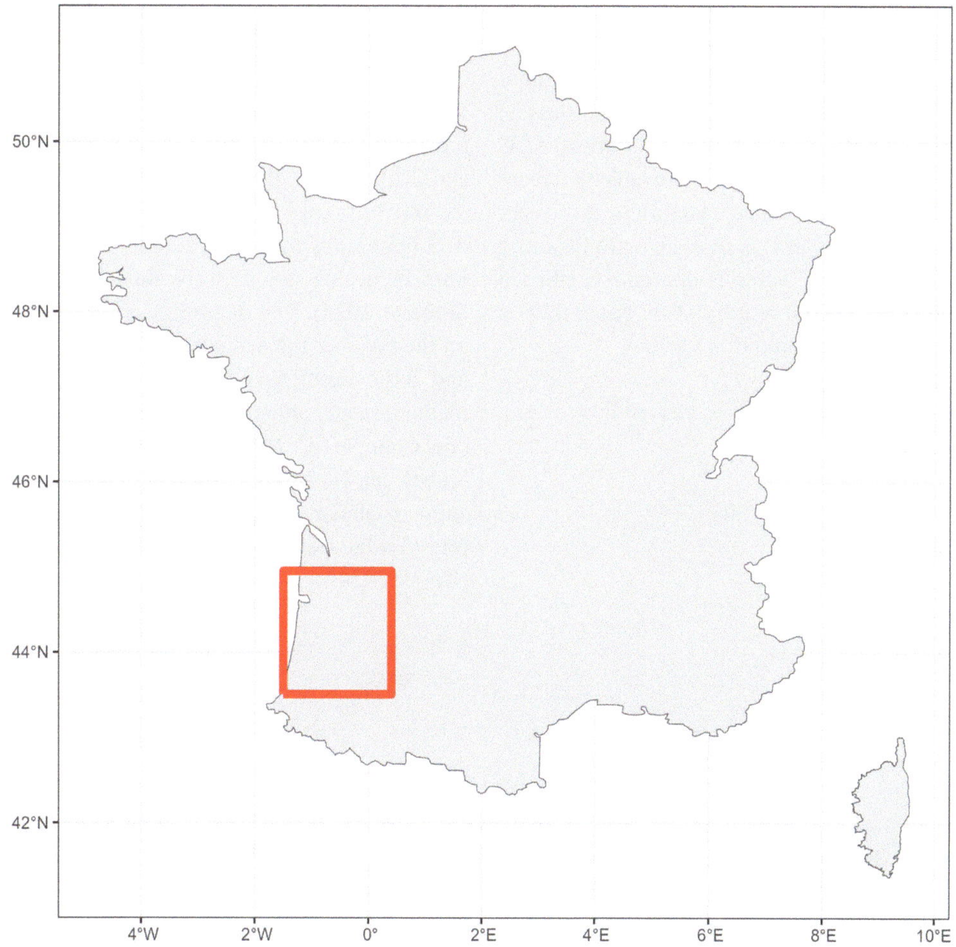

Fig. 10.3 Map of France with a polygon indicating the study area with forest damage caused by the Klaus windstorm in January 2009

vidual tree and the structure and size of the forest (Probst et al. 2019b). Typically, the search for optimal values is performed by testing different candidate values using, for example, **k-fold cross-validation** and finding those that yield the highest accuracy (Probst et al. 2019a). For an RF, we can tune the ***number of trees***, the ***mtry*** parameter, and the ***minimum node size*** (min_n). Many trees is desirable because their results are then averaged (aggregated) to obtain the final result. When a single tree is built, a random sample of m predictors, *mtry*, is taken at each split in the tree (Pawlik and Harrison 2022). For classification, these values should be $mtry \approx \sqrt{p}$, where p is the number of predictor variables (Kuhn and Johnson 2013; Probst et al. 2019b). Lower *mtry* leads to less correlated trees, supporting stability during the aggregation of results; however, low values of *mtry* can lead to worse-than-average performance at the tree level (Probst et al. 2019b). This is because during the random selection of potential predictors, nonimportant variables are chosen. The minimum node size is the minimum number of observations at a terminal node. A low value of *min_n* resultes in a large tree depth, i.e., splitting is performed many times until the terminal node is reached (Probst et al. 2019b). The optimal hyperparameters are searched and selected on the basis of model performance, as evaluated on the basis of the following metrics: **accuracy**, receiver operating characteristic (ROC) curve, **sensitivity**, and **specificity**. Additionally, the basic confusion matrix rate, or misclassification rate, is commonly used (Chicco et al. 2021). **Accuracy** is the proportion of all correct classifications, positive and negative, calculated on the basis of the total number of observations. The ROC curve is a graphical method for evaluating binary classification models (Hajian-Tilaki 2013; Nahm 2022), whereas the area under the ROC curve (AUC) reflects the entire scope of the ROC curve and is widely used as a measure of model performance (Hajian-Tilaki 2013). The ROC curve reflects the sensitivity (i.e., recall, the number of samples in the true positive class divided by the total number of positive results) and specificity (the number of samples in the true negative class divided by the sum of false positives and true negatives) over a continuum of different event thresholds and is a measure of the probability of correctly identifying a positive signal in a noise background (Hanley and McNeil 1982; Kuhn and Johnson 2013). The following interpretation of AUC values was adopted for the present task: 0.5 (random discrimination between binary classes), 0.5–0.7 (weak discrimination), 0.7–0.8 (acceptable discrimination), 0.8–0.9 (excellent discrimination), 0.9-1.0 (outstanding discrimination), and 1.0 (perfect discrimination between classes) (Pawlik et al. 2022). Note that this classification of AUC values is rule of thumb, and the AUC is prevalence dependent; therefore, in cases with unbalanced datasets, it is easier to obtain high AUC values than it is in cases with balanced datasets. In addition, the AUC value is domain specific; in species distribution models, 0.8 is fair, 0.9 is good, 0.95 is excellent, and the level of interpretation varies.

```
#In this project, we first define the study area.


pacman::p_load("sf", "rgeoboundaries",
"ggplot2")


library(sf)
library(rgeoboundaries)

#Select a vector layer with a border of France.
france <- gb_adm0("France")

#Definition of the study area with forest damage
caused by windstorm Klaus in 2009.


forwind_df <- data.frame(long = c(-1.5053621,
0.3993588, 0.3993588, -1.5053621, -1.5053621),
 lat = c(43.50881, 43.50881, 44.95153,
44.95153, 43.50881))

#Transform points into a polygon.
forwind_polygon <- forwind_df |>
 st_as_sf(coords = c("long", "lat"), crs =
4326) |>
 st_combine() |>
 st_cast("LINESTRING")

library(ggplot2)


#Plot the map.
ggplot()+
  geom_sf(data = france)+
  geom_sf(data = forwind_polygon, col ="red",
linewidth = 2)+
  theme_bw()

#Save the map.
#ggsave("france_study_area.png")
```

10.2 R in Practice: Forest Damage Due to Strong Winds

With the available dataset, we can start to train and build a predictive model. We test three R packages: the **caret** package (Kuhn 2008), the **iml** package (Molnar et al. 2018), and the **mlr3** package (Lang et al. 2019). An exploration of the data before any modeling attempt is recommended. We can implement this step with the **flextable** package (Gohel and Skintzos 2024). The dataset is stratified into two categories on the basis of the categorical response variable "damage", and basic descriptive statistics are calculated: the mean, median, range, and missing values (Table 10.1). This is just one example of data exploration, and several other techniques can be applied (see Sect. 1.3). Other packages that support tabular data and statistical reporting include **tableone** (Yoshida and Bartel 2022) and the **reporter** package (Bosak 2024).

```
#This part relates to the initial data
exploration and modeling stage.


pacman::p_load("flextable")


library(rappliES)


#Import the dataset with information on forest
damage caused by the windstorm Klaus in 2009
and potential predictors from the rappliES
package.


dataset_klaus <- klaus


#Checking variable names and their number.
names(dataset_klaus)
length(names(dataset_klaus))
head(dataset_klaus)

#With the flextable package, you can get basic
characteristics of the dataset.


library(flextable)


#Start with a univariate statistical analysis.
summtab <- summarizor(dataset_klaus[,-1], by =
"damage")
```

Table 10.1 The list of variables used for training the classification model by applying the random forest method. The table was prepared with the **flextable** package

		Dam (N=2,973)	No_dam (N=2,973)
elevation	Mean (SD)	77.4 (30.0)	71.7 (41.2)
	Median (IQR)	75.7 (38.4)	66.7 (57.6)
	Range	10.3–169.9	−1.2–228.3
	Missing		33 (1.1%)
slope	Mean (SD)	0.0 (0.0)	0.0 (0.0)
	Median (IQR)	0.0 (0.0)	0.0 (0.0)
	Range	0.0–0.2	0.0–0.3
	Missing		35 (1.2%)
roughness	Mean (SD)	2.1 (1.6)	2.5 (2.4)
	Median (IQR)	1.7 (1.7)	1.8 (2.2)
	Range	0.1–15.9	0.0–22.1
	Missing		33 (1.1%)
profile_curv	Mean (SD)	0.0 (0.0)	−0.0 (0.0)
	Median (IQR)	0.0 (0.0)	−0.0 (0.0)
	Range	-0.0–0.0	−0.0–0.0
	Missing		68 (2.3%)
plane_curv	Mean (SD)	0.0 (0.0)	0.0 (0.0)
	Median (IQR)	0.0 (0.0)	0.0 (0.0)
	Range	−0.0–0.0	−0.0–0.0
	Missing		68 (2.3%)
wind_speed	Mean (SD)	37.3 (2.7)	35.8 (3.9)
	Median (IQR)	37.2 (3.9)	35.1 (5.6)
	Range	31.0–45.0	27.3–47.7
	Missing	1 (0.0%)	121 (4.1%)
twi	Mean (SD)	7.2 (0.9)	7.2 (1.3)
	Median (IQR)	7.2 (1.2)	7.3 (1.7)
	Range	3.8–10.4	3.2–11.4
	Missing		33 (1.1%)
wind_exp	Mean (SD)	1.0 (0.0)	1.0 (0.1)
	Median (IQR)	1.0 (0.1)	1.0 (0.1)
	Range	0.9–1.2	0.8–1.2
	Missing		33 (1.1%)
coastline_dist	Mean (SD)	44,613.4 (25,246.5)	49,647.3 (32,359.5)
	Median (IQR)	42,000.0 (43,000.0)	46,000.0 (51,000.0)
	Range	3,000.0–110,000.0	1,000.0–131,000.0
	Missing	1 (0.0%)	41 (1.4%)
wind_track_dist	Mean (SD)	135,928.0 (26,080.7)	112,196.5 (43,357.1)
	Median (IQR)	137,000.0 (39,000.0)	111,000.0 (75,000.0)
	Range	56,000.0–197,000.0	23,000.0–201,000.0
	Missing	1 (0.0%)	41 (1.4%)
clay_content	Mean (SD)	18.3 (3.0)	19.6 (3.6)
	Median (IQR)	18.3 (4.2)	19.8 (4.9)
	Range	10.2–27.3	9.2–30.8
	Missing	1 (0.0%)	63 (2.1%)
lai	Mean (SD)	2.5 (1.2)	2.6 (3.0)
	Median (IQR)	2.3 (1.6)	2.0 (1.7)
	Range	0.2–15.6	0.3–25.4
	Missing		42 (1.4%)
ndvi	Mean (SD)	0.8 (0.1)	0.7 (0.1)
	Median (IQR)	0.8 (0.1)	0.7 (0.1)
	Range	0.4–0.9	−0.3–0.9
	Missing		35 (1.2%)

```
#Create flextable bindings from an object.
ft <- as_flextable(summtab)

#Change the header color.
ft <- color(ft, color = "orange", part =
"header")
ft

#Save the table as an image.
#save_as_image(ft, "flextable.png")
```

Now, it is the time to prepare the dataset for modeling. The **caret** (short for **C**lassification **A**nd **Re**gression **T**raining) package contains several functions that help prepare the data. First, we split the dataset into training and test sets using the ***createDataPartition()*** function. A total of 70% of the observations that are randomly sampled are used for the training procedure, and 30% are used for model testing (Kuhn 2019). The proportion of observations in the training and test datasets is arbitrary and can be set differently to obtain the best possible results. For instance, a 40% split rate can be tested. It is important to test the model's predictive power with an independent validation set that was not part of the model training dataset (it was "unseen" by the model). This approach helps us evaluate the model effectively. Note that the data for the response variable are balanced, i.e., the proportions of positive and negative cases are equal in both the training and validation sets (Table 10.1). Another step is to remove highly correlated variables. With the function ***findCorrelation()***, we identify and remove all variables with correlation coefficients greater than 0.75. The function calculates the correlation matrix and returns a vector of integers indicating columns that will be removed to reduce pairwise correlations. After splitting and removing highly correlated variables, we can define the arguments that control the training procedure. For this step, the ***trainControl()*** function arguments need to be selected.

```
library(caret)  #package offering functions for
model training and plotting


#Calculate correlations and find values > 0.75.
klaus_cor <- cor(dataset_klaus[,3:15], use =
'complete.obs')
highCorr_klaus <- findCorrelation(klaus_cor,
0.75, names = TRUE)
highCorr_klaus

#Remove highly correlated variables from
training and test sets.
dataset_klaus_reduced <- dataset_klaus |>
 dplyr::select(-highCorr_klaus[1],
 -highCorr_klaus[2],
 -highCorr_klaus[3],
 -highCorr_klaus[4])

#Set the seed of the pseudorandom number
generator. It guarantees the reproducibility of
the results presented in this book. Because the
splitting is random, we need some control over
the process.


set.seed(111)


#Data partition into the training and test
sets.
klaus_partition <- createDataPartition(dataset_
klaus_reduced$damage,
 p = 0.7, list = FALSE)

train_klaus <- dataset_klaus_reduced[klaus_
partition,]
test_klaus <- dataset_klaus_reduced[-klaus_
partition,]

#Still, there are some NA values that need to
be removed.


summary(train_klaus_reduced)
summary(test_klaus_reduced)

#We also remove the first column with the row
index.


train_full <- train_klaus_reduced[complete.
cases(train_klaus_reduced),-1]
test_full <- test_klaus_reduced[complete.
cases(test_klaus_reduced),-1]

#In this part, we set the trainControl()
function.


#Random Forest modeling control parameters.
fit_Control_RF <- trainControl(method =
"repeatedcv",
                              number = 5,
                              repeats = 5,
                              classProbs =
TRUE,
                              summaryFunction
= twoClassSummary,
                              savePredictions
= TRUE)
```

```
pacman::p_load("tidyselect", "pROC", "vip",
"ConfusionTableR", "plotROC", "gt")


library(ggplot2)
library(tidyselect) #a backend for the
selecting functions of the 'tidyverse'
library(pROC) #package for ROC curves display
and analyze

#Hyperparameter tuning.
tgrid <- expand.grid(
 mtry = c(2,4,6),          #number of variables
randomly selected at each split
 splitrule = "gini",      #splitting criterion
(classification problem)
 min.node.size = c(10, 20, 30, 40) #minimum
size of terminal nodes
)

#Random Forest modeling.
set.seed(123) # for reproducibility

rf_model_klaus <- caret::train(
  damage ~ .,                  # formula with
the response variable
  data = train_full,           # training
dataset
  method = "ranger",           # random Forest
implementation in caret
  trControl = fit_Control_RF,   # control
parameters for resampling
  metric = "ROC",              # optimization
metric
  preProcess = c("center", "scale"),
  num.trees = 200,             # number of
trees
  tuneGrid = tgrid,            # grid of
hyperparameters
  importance = 'impurity')

#Results for each combination of mtry and min.
node.size.
rf_model_klaus$results
#The best result.
rf_model_klaus$bestTune

library(ConfusionTableR) #tools for confusion
matrix plotting


#RF model applied to the test dataset. The type
argument was set to "raw" the result will be
predicted classes of the response variable.
test_pred_rf <- predict(rf_model_klaus, newdata
= test_full,
 type = 'raw')

#With this prediction, the confusion matrix can
be displayed and examined.
caret::confusionMatrix(test_pred_rf, as.
factor(test_full[,1]), positive = 'dam')

#In this way, only the matrix can be displayed.
as.matrix(caret::confusionMatrix(test_pred_rf,
as.factor(test_full[,1]), positive = 'dam'))

#We combine the result of the predict()
function with the test dataset.
predictedRF <- cbind(data.frame(class_preds =
test_pred_rf, test_full))

#Plotting of the confusion matrix.
#png("conf_mx.png", width = 600, height = 500)
ConfusionTableR::binary_visualiseR(train_labels
= predictedRF$class_preds, truth_labels =
as.factor(predictedRF$damage), class_label1 =
"damage", class_label2 = "no_damage",
                                    quadrant_
col1 = "#28ACB4",
                                    quadrant_
col2 = "#4397D2",
                                    custom_title
= "Forest damage confusion matrix",
                                    text_col =
"black"
                                    )
#dev.off()

#We need to calculate predictions as
probabilities for the ROC curve and the AUC. In
this case, the type argument of the predict()
function is set to "prob".


test_pred_rf_prob <- predict(rf_model_klaus,
newdata = test_full,
                              type = 'prob')

#ROC (Receiver Operating Characteristics)
curve.


rf_auc <- pROC::roc(test_full$damage, test_
pred_rf_prob$dam)


#png("roc_rf.png")
plot(rf_auc, grid=TRUE, print.auc=TRUE,
col='blue',
     print.auc.x=0.9, print.auc.y=0.5)
#dev.off()
```

```
#The predictions need to be in a data frame
form to use the ggplot2 and the plotROC
package.


rf_auc_df <- data.frame(obs = rf_model_
klaus$pred$obs,

dam = rf_model_klaus$pred$dam,

roc = rf_model_klaus$results$ROC,
                          model = "RF
model")

library(plotROC) #plotting and interactive
tools for ROC curves


#Plotting.
ggplot()+
  geom_roc(rf_auc_df,
         mapping = aes(m = dam, d = factor(obs,
levels = c('no_dam','dam'))),
         n.cuts=0, color = 'blue',
         increasing = FALSE)+
  theme_bw()

#Saving the plot.
#ggsave("roc_rf_geom_roc_function.png")
```

For model training, five-fold cross-validation was selected and repeated 5 times. Another common k-fold validation option is ten-fold cross-validation. For hyperparameter tuning, we set up a search grid for three potential *mtry* values (2, 4, and 6) and *min_n* (10, 20, 30, and 40). As a result, after the training procedure, 12 combinations of these values were obtained with the corresponding evaluation metrics, and from these results, the best combination was selected. We refer to model components using the **$** operator in the RStudio console: *model$results* and *model$bestTune*. Model performance is evaluated on the basis of the **ROC**, **sensitivity** (the proportion of correctly classified positive observations), and **specificity** (the proportion of correctly classified negative observations) (Robin et al. 2011). In our case, the best *mtry* value is 4, and *min_n* is 30. Note that these parameters can differ from the results obtained in completely new sessions because the training procedure is implemented with different randomly selected observations. With these optimal parameters, the accuracy of the random forest model reached 0.781 (Fig. 10.4), and the AUC was 0.859 (Figs. 10.5 and 10.6). We want to mitigate model overfitting or when the model works well for the training dataset but performs poorly

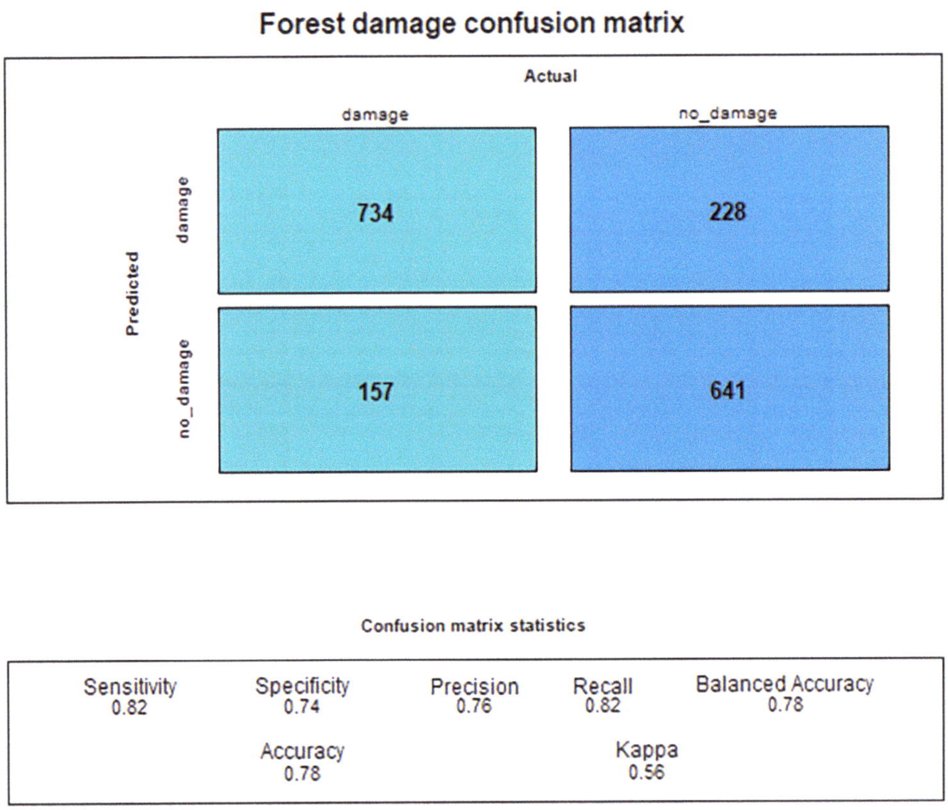

Fig. 10.4 Confusion matrix showing the performance of the random forest model for the test set. The figure was prepared with the **ConfusionTable R** package (Hutson 2021)

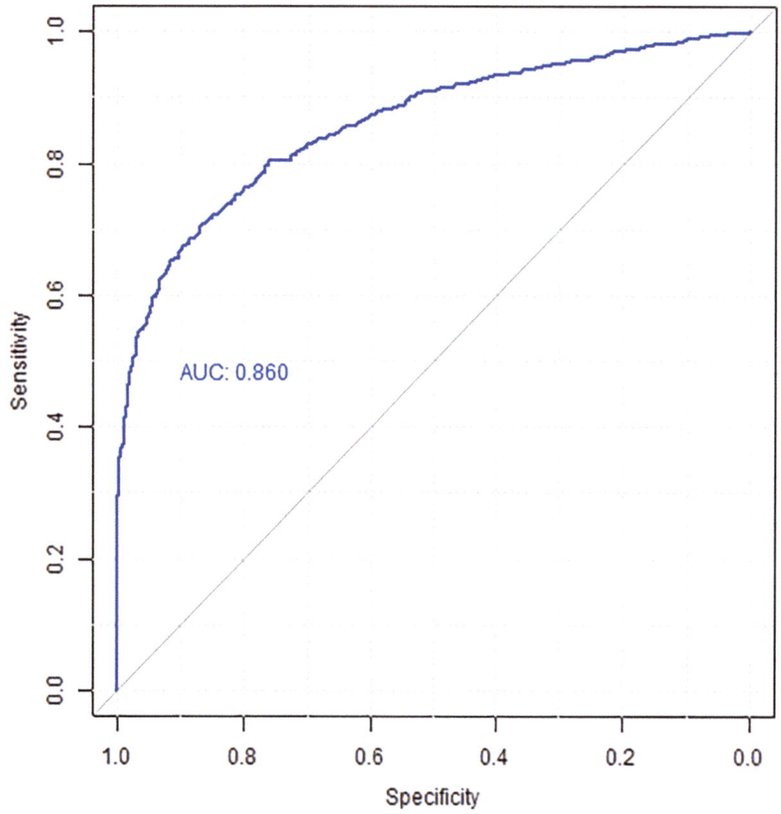

Fig. 10.5 The ROC curve for the best random forest model applied to the test set

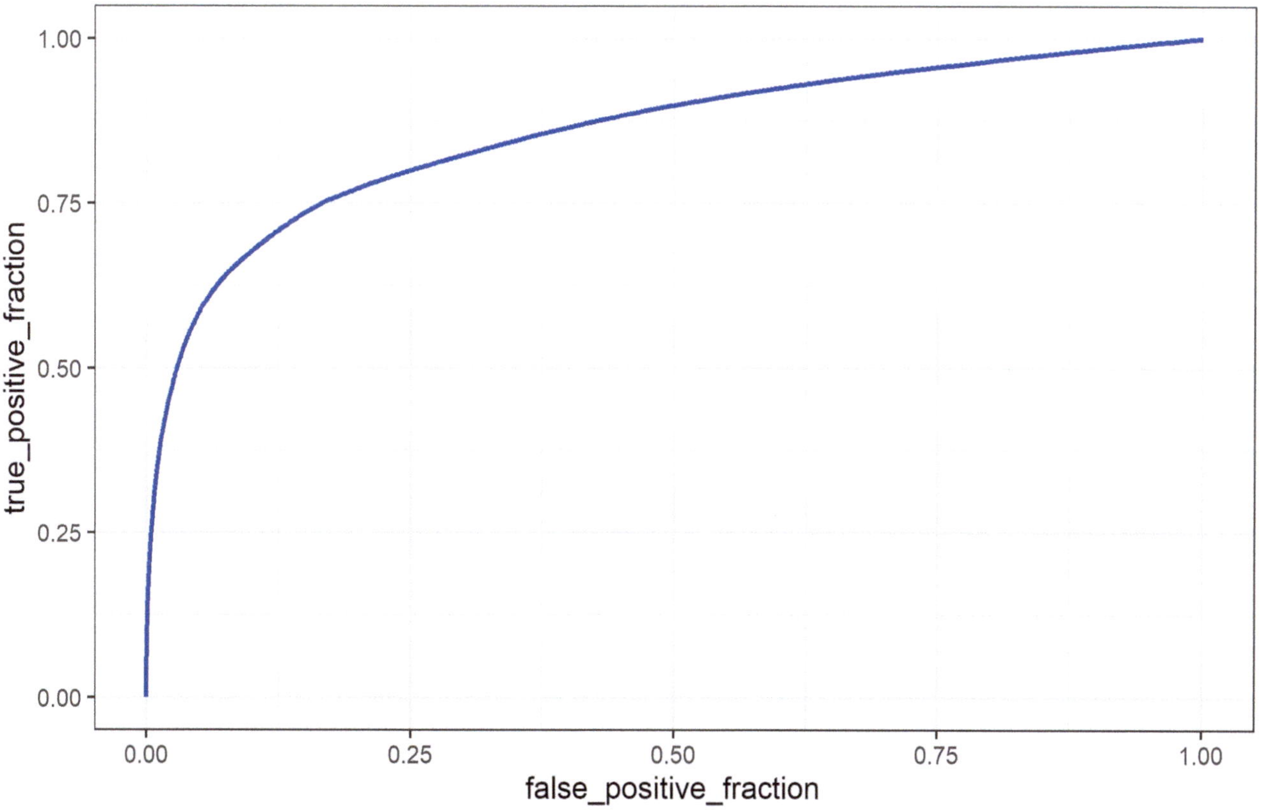

Fig. 10.6 An alternative way of plotting the ROC curve is to use the ***geom_roc()*** function from the **plotROC** package (Sachs 2017). The true positive fraction (Y-axis) represents sensitivity (the proportion of correctly classified positive observations), and the false positive fraction (X-axis) represents specificity (the proportion of correctly classified negative observations)

when it is applied to new data. This can occur if the training dataset is very small, which limits the likelihood of learning general patterns hidden in the data. Increasing the number of observations in the training dataset can aid in building a better model, but a larger test set must also be created to evaluate the resulting predictive model. On the basis of the confusion matrix created using the **ConfusionTableR** package (Hutson 2021), 157 (18%) actual forest damage observations were incorrectly labeled "no_damage". Eighty-two percent of the observations were correctly classified as "damage", and this metric is referred to as sensitivity or recall. **Balanced accuracy** is the arithmetic mean of sensitivity and specificity (Chicco et al. 2021). The kappa coefficient expresses the agreement between two nominal classifications, which are based on the same categories (Delgado and Tibau 2019; Chicco et al. 2021).

We know that the model provides high accuracy, but how can we rank the predictors in terms of their importance? The question can be answered with a variable importance plot (Fig. 10.7). For this step, we use the **vip** package (Greenwell and Boehmke 2020). The most important predictor is the distance of the forest damage area from the track of the windstorm Klaus (Table 10.2). Other important predictors are the NDVI, the wind exposure index (wind_exp), and the distance from the Atlantic Ocean coastline.

```
pacman::p_load("vip", "gt")


#At this stage, we explore the importance of
variables (predictors). Their predictive power
is checked against other potential predictors
(independent variables) considered during the
model training procedure.


library(vip)
library(gt)
```

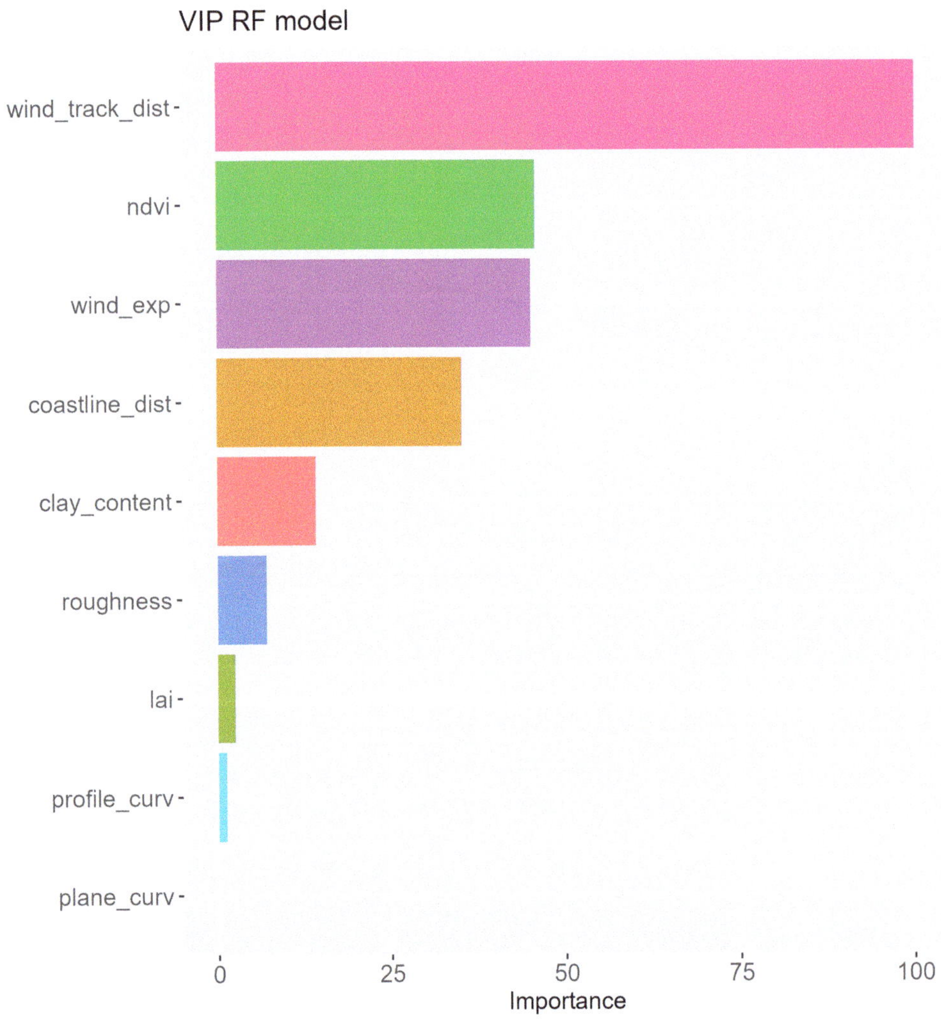

Fig. 10.7 Variable importance plot for the best RF model

Table 10.2 Importance of the predictors for the best RF model

predictor	Overall
wind_track_dist	100.0000000
wind_exp	46.6077145
ndvi	42.7835415
coastline_dist	30.6383800
clay_content	19.2780402
slope	7.7670229
lai	3.9515952
profile_curv	0.8143756
plane_curv	0.0000000

```
#Calculate variable importance.
vi_rf <- varImp(rf_model_klaus)

#Show it as a table using the gt package.


vi_tbl <- vi_rf$importance |>
  tibble::rownames_to_column(var = "predictor")
|>
  dplyr::arrange(desc(Overall)) |>
  gt() |>
  tab_style(style = cell_text(weight = "bold"),
              locations = cells_body(columns
= "Overall")) |>
  data_color(apply_to = "text", palette =
"Spectral", columns = "Overall")
vi_tbl

#Save the table as a plot.
#gtsave(vi_tbl, "vip_gt.png")

vi_rf_plot <- vip(rf_model_klaus,
                 mapping = aes(fill=Variable))+
  ggtitle('VIP RF model')+
  theme
(axis.text = element_text(size=12),
        plot.title = element_text(size=14),
        axis.title.x = element_text(size=12))
vi_rf_plot

#Save the plot.
#ggsave("vip_rf_model.png")
```

We look closely at the impacts of individual predictors using a partial dependence plot (PDP), which is a global model-agnostic tool. This plot provides information about the average behavior of a ML model (Molnar 2022). With a PDP, the expected prediction is checked when all other features are marginalized (Molnar 2022). The result indicates whether the relationship is linear, monotonic, or more complex. We run ***FeatureEffects()*** with the **iml** package (Molnar et al. 2018), setting the method argument to "pdp". We learned from the VIP plot that the windstorm track distance is the most important predictor, and Fig. 10.8 shows that this effect is particularly strong for distances close to 150 km. The damage probability increases with increasing NDVI >0.5. This finding means that healthy trees were the trees most damaged during the windstorm, and we can exclude factors with a negative impact on forest stand conditions, such as air pollution or bark beetle outbreaks, that might have made the trees more prone to wind damage.

```
pacman::p_load("iml")


library(iml)


#Random forest model.
x_rf = rf_model_klaus$trainingData[which(names
(rf_model_klaus$trainingData) != "dam")]

#Calculate prediction probabilities.
model_rf = Predictor$new(rf_model_klaus, data =
x_rf[,-1], y = rf_model_klaus$trainingData$.
outcome,
 type = "prob", class = "dam")

#Calculate feature effects.
effect_rf = FeatureEffects$new(predictor = model_
rf, method = "pdp", grid.size = 30)

#Plot the PDP.
rf_pdp_plots <- plot(effect_rf, fixed_y = FALSE)

#Save the plot.
#png("pdp_plot.png", width = 800, height = 600)
rf_pdp_plots
#dev.off()
```

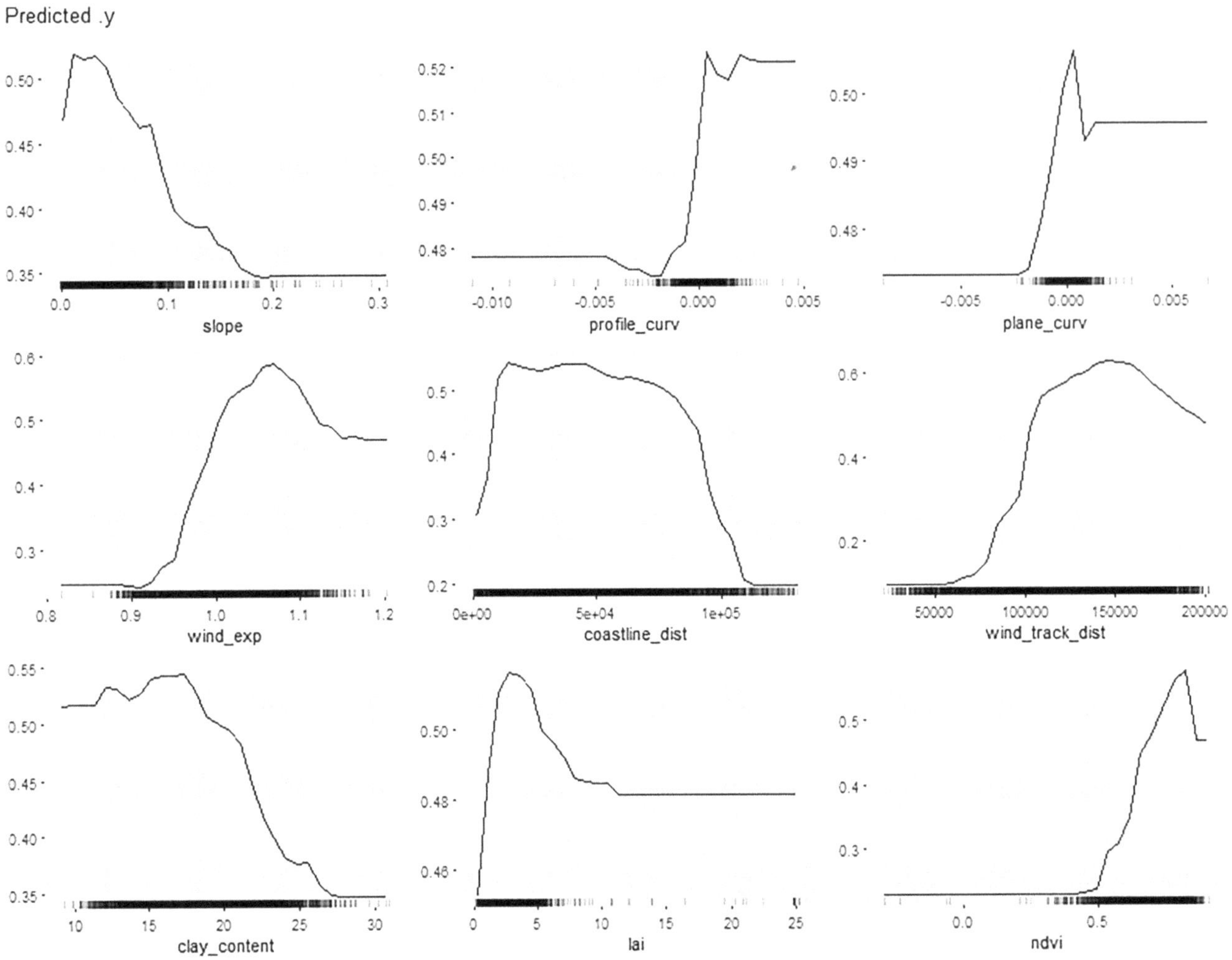

Fig. 10.8 Partial dependence plot of the impact of predictors on the probability of forest damage

References

Badoux A, Andres N, Techel F, Hegg C (2016) Natural hazard fatalities in Switzerland from 1946 to 2015. Nat Hazards Earth Syst Sci 16:2747–2768. https://doi.org/10.5194/nhess-16-2747-2016

Benestad RE, Mezghani A, Parding KM, et al (2024) ESD: climate analysis and empirical-statistical downscaling (ESD) package for monthly and daily data

Beniston M, Stephenson DB (2004) Extreme climatic events and their evolution under changing climatic conditions. Glob Planet Chang 44:1–9. https://doi.org/10.1016/j.gloplacha.2004.06.001

Bosak D (2024) Reporter: creates statistical reports

Breiman L (2001) Random forests. Mach Learn 45:5–32

Brönnimann S, Martius O, von Waldow H et al (2012) Extreme winds at Northern mid-latitudes since 1871. Meteorol Z 21:013–027

Charlton-Perez AJ, Vukadinovic Greetham D, Hemingway R (2019) Storm naming and forecast communication: a case study of storm Doris. Meteorol Appl 26:682–697. https://doi.org/10.1002/met.1794

Chaudhary MT, Piracha A (2021) Natural disasters—origins, impacts, management. Encyclopedia 1:1101–1131. https://doi.org/10.3390/encyclopedia1040084

Chicco D, Tötsch N, Jurman G (2021) The Matthews correlation coefficient (MCC) is more reliable than balanced accuracy bookmaker informedness and markedness in two-class confusion matrix evaluation Abstract BioData Mining 14(1) https://doi.org/10.1186/s13040-021-00244-z

Dawkins LC, Stephenson DB, Lockwood JF, Maisey PE (2016) The 21st century decline in damaging European windstorms. Nat Hazards Earth Syst Sci 16:1999–2007. https://doi.org/10.5194/nhess-16-1999-2016

Delgado R, Tibau X-A (2019) Why Cohen's kappa should be avoided as performance measure in classification. PLoS One 14:e0222916. https://doi.org/10.1371/journal.pone.0222916

Donat MG, Leckebusch GC, Wild S, Ulbrich U (2011) Future changes in European winter storm losses and extreme wind speeds inferred from GCM and RCM multi-model simulations. Nat Hazards Earth Syst Sci 11:1351–1370. https://doi.org/10.5194/nhess-11-1351-2011

Donatti CI, Nicholas K, Fedele G et al (2024) Global hotspots of climate-related disasters. Int J Disaster Risk Reduct 108:104488. https://doi.org/10.1016/j.ijdrr.2024.104488

Forzieri G, Pecchi M, Girardello M et al (2020) A spatially explicit database of wind disturbances in European forests over the period 2000–2018. Earth Syst Sci Data 12:257–276. https://doi.org/10.5194/essd-12-257-2020

Gardiner B, Blennow K, Carnus JM et al (2010) Destructive storms in European forests: past and forthcoming impacts. In: Final report to European Commission—DG environment. European Forest Institute, Joensuu

Gardiner B, Schuck A, Schelhaas M-J et al (2013) Living with storm damage to forests. European Forest Institute

Godziek J, Pawlik Ł (2023) Indicators of wind-driven forest disturbances—pit-mound topography, its automatic detection and significance. Catena 221:106757. https://doi.org/10.1016/j.catena.2022.106757

Gohel D, Skintzos P (2024) Flextable: functions for tabular reporting

Greenwell B, Boehmke B (2020) Variable importance plots—an introduction to the vip package. R J 12:343–366

Hajian-Tilaki K (2013) Receiver operating characteristic (ROC) curve analysis for medical diagnostic test evaluation. Caspian J Intern Med 4:627–635

Hanley JA, McNeil BJ (1982) The meaning and use of the area under a receiver operating characteristic (ROC) curve. Radiology 143:29–36. https://doi.org/10.1148/radiology.143.1.7063747

Hutson G (2021) ConfusionTableR: confusion matrix toolset

Kuhn M (2008) Building predictive models in R using the caret package. J Stat Softw 28:1–26. https://doi.org/10.18637/jss.v028.i05

Kuhn M (2019) The caret package

Kuhn M, Johnson K (2013) Applied predictive modeling. Springer, New York

Lang M, Binder M, Richter J et al (2019) mlr3: a modern object-oriented machine learning framework in R. J Open Sour Software 4:1903. https://doi.org/10.21105/joss.01903

Lesmeister C, Chinnamgari SK (2019) Advanced machine learning with R. Packt, Birmingham, Mumbai

Lindtroth A, Lagergren F, Grelle A et al (2009) Storms can cause Europe-wide reduction in forest carbon sink. Glob Chang Biol 15:346–355. https://doi.org/10.1111/j.1365-2486.2008.01719.x

Machado Nunes Romeiro J, Eid T, Antón-Fernández C et al (2022) Natural disturbances risks in European boreal and temperate forests and their links to climate change—a review of modelling approaches. For Ecol Manag 509:120071. https://doi.org/10.1016/j.foreco.2022.120071

McPhillips LE, Chang H, Chester MV et al (2018) Defining extreme events: a cross-disciplinary review. Earth's Future 6:441–455. https://doi.org/10.1002/2017EF000686

Molnar C (2022) Interpretable machine learning. A guide for making black Box models explainable, 2nd edn. Mucbook Clubhouse, München

Molnar C, Bischl B, Casalicchio G (2018) iml: an R package for interpretable machine learning. J Open Source Soft 3:786. https://doi.org/10.21105/joss.00786

Nahm FS (2022) Receiver operating characteristic curve: overview and practical use for clinicians. Korean J Anesthesiol 75:25–36. https://doi.org/10.4097/kja.21209

Neu U, Akperov MG, Bellenbaum N et al (2013) IMILAST: a community effort to Intercompare extratropical cyclone detection and tracking algorithms. Bull Am Meteorol Soc 94:529–547. https://doi.org/10.1175/BAMS-D-11-00154.1

Pawlik Ł, Harrison SP (2022) Modelling and prediciton of wind damage in forest ecosystems of the Sudety Mountains, SW Poland. Sci Total Environ 815:151972. https://doi.org/10.1016/j.scitotenv.2021.151972

Pawlik Ł, Migoń P, Szymanowski M (2016) Local-and regional-scale biomorphodynamics due to tree uprooting in semi-natural and managed montane forests of the Sudetes Mountains, Central Europe. Earth Surf Process Landf 41:1250–1265. https://doi.org/10.1002/esp.3950

Pawlik Ł, Godziek J, Zawolik Ł (2022) Forest damage by extra-tropical cyclone Klaus—modeling and prediction. Forests 13:1991. https://doi.org/10.3390/f13121991

Probst P, Boulesteix A-L (2017) To tune of not to tune the number of trees in random forest?

Probst P, Boulesteix A-L, Bischl B (2019a) Tunability: importance of Hyperparameters of machine learning algorithms. J Mach Learn Res 20:1–32

Probst P, Wright MN, Boulesteix A-L (2019b) Hyperparameters and tuning strategies for random forest. WIREs Data Min Knowl Discov 9:e1301. https://doi.org/10.1002/widm.1301

Roberts JF, Champion AJ, Dawkins LC et al (2014) The XWS open access catalogue of extreme European windstorms from 1979 to 2012. Nat Hazards Earth Syst Sci 14:2487–2501. https://doi.org/10.5194/nhess-14-2487-2014

Robin X, Turck N, Hainard A et al (2011) pROC: an open-source package for R and S+ to analyze and compare ROC curves. BMC Bioinform 12:77. https://doi.org/10.1186/1471-2105-12-77

Ruosteenoja K, Vihma T, Venäläinen A (2019) Projected changes in European and North Atlantic seasonal wind climate derived from CMIP5 simulations. J Clim 32:6467–6490. https://doi.org/10.1175/JCLI-D-19-0023.1

Sachs MC (2017) plotROC: a tool for plotting ROC curves. J Stat Softw 79:1–19. https://doi.org/10.18637/jss.v079.c02

Schelhaas M-J, Nabuurs G-J, Schuck A (2003) Natural disturbances in the European forests in the 19th and 20th centuries. Glob Chang Biol 9:1620–1633. https://doi.org/10.1046/j.1365-2486.2003.00684.x

Schwierz C, Köllner-Heck P, Mutter EZ et al (2010) Modelling European winter wind storm losses in current and future climate. Climate Change 101:485–514. https://doi.org/10.1007/s10584-009-9712-1

Seidl R, Thom D, Kautz M et al (2017) Forest disturbances under climate change. Nat Clim Chang 7:395–402. https://doi.org/10.1038/NCLIMATE3303

Senf C, Pflugmacher D, Zhiqiang Y et al (2018) Canopy mortality has doubled in Europe's temperate forests over the last three decades. Nat Commun 9:4978. https://doi.org/10.1038/s41467-018-07539-6

Smith MD (2011) An ecological perspective on extreme climatic events: a synthetic definition and framework to guide future research. J Ecol 99:656–663. https://doi.org/10.1111/j.1365-2745.2011.01798.x

Spinoni J, Formetta G, Mentaschi L et al (2020) Global warming and windstorm impacts in the EU. Publications Office of the European Union

Stewart M, Carleton WC, Groucutt HS (2022) Extreme events in biological, societal, and earth sciences: a systematic review of the literature. Front Earth Sci 10:1. https://doi.org/10.3389/feart.2022.786829

Surjalal Sharma A, Baker DN, Bhattacharyya A et al (2012) Complexity and extreme events in geosciences: an overview. In: Extreme events and natural hazards: the complexity perspective geophysical monograph. American Geophysical Union, pp 1–17

Svensen H (2009) The end of nigh. In: A history of natural disasters. Reaktion Books, London

Yoshida K, Bartel A (2022) Tableone: create "table 1" to describe baseline characteristics with or without propensity score weights

Knowledge in the Cloud: LiDAR Point Cloud Data

11

Trephina Gorge Nature Park, Central Australia, 2022, photo: Ł. Pawlik

Supplementary Information The online version contains supplementary material available at https://doi.org/10.1007/978-3-031-89673-6_11.

© The Author(s), under exclusive license to Springer Nature Switzerland AG 2025

Ł. Pawlik, *R Applications in Earth Sciences*, Springer Textbooks in Earth Sciences, Geography and Environment,
https://doi.org/10.1007/978-3-031-89673-6_11

Advances in airborne LiDAR systems make it possible to acquire high-quality terrain data in terms of accuracy and density.

(Xiaoye Liu 2008)

With thousands of points registered from laser beams per second LiDAR technology is currently the most powerful tool for Earth surface mapping. An active remote sensing system uses a laser to measure the distance to a target (Wulder et al. 2008). This technology emits intense, focused beams of light and measures the time of their reflections (NOAA 2012), collecting information about the object's position, including X, Y, and Z values (longitude, latitude, and elevation). The generated product is a point cloud, which is a set of highly accurate georeferenced elevation points (NOAA 2012). Because of the range of aircraft and drone missions and the speed of data collection, it is now possible to generate large high-quality datasets at the regional scale of analysis. With several advantages, the use of LiDAR technology also requires overcoming several disadvantages, starting from the high cost of data collection and commercial software and extending to the large volume of data that must be processed and stored, which requires a substantial amount of memory storage. In addition, the processing of the data is complex and requires specialized knowledge and sometimes programming skills.

The growing demand for high-precision data has led to further advances in remote sensing techniques and the collection of massive amounts of data using automatic and fast methods. Precise information regarding the ground surface, man-made objects, or vegetation is now required in several branches of science and by public administrations and companies. For example, flood hazard and risk maps cannot be produced without information on the topography and elevation of river valley floors, with precision measured in centimeters, not meters, and with information on the area of river floodplains covered by buildings and infrastructure (Dysarz 2020). Both features of the ground surface properties can be obtained through airborne scanning, point cloud processing, and modeling. The final product, a digital terrain model, can be used for other applications. Another example of the increasing use of high-resolution data derived from LiDAR point clouds is vegetation classification and characterization. LiDAR-based models of vegetation cover, and recently, even individual tree detection, have become popular in forest ecology and forestry (Zhen et al. 2016; Murtiyoso et al. 2024). New methods support the detection of dead tree trunks (Heinaro et al. 2021), fallen tree root plates (Godziek 2024), and pit-mound topography, even in areas with dense forest cover (Godziek and Pawlik 2023).

11.1 Description of the Data and Project Outline

In this section, we focus on the spectacular landscape of the Colorado River in Utah in the United States (Fig. 11.1). The subject of our analysis is just a slice of the spatial data covering Dead Horse Point State Park. The state park is located between Arches National Park and Canyonlands National Park. On the basis of the LiDAR point cloud, we show the topography of a part of the area where the Colorado River has cut deeply into lithified deposits of various origins and ages (Fig. 11.2). The geology of this cross-section varies from shallow marine deposits of Pennsylvanian age (323–299 million years ago) at the bottom to aeolian sandstones and river deposits of Jurassic age (199–145 million years ago) at the top of the cliff-like valley sides (Doelling and Chidsey Jr. 2000). The data can be searched and downloaded from the USGS 3DEP LidarExplorer (https://apps.nationalmap.gov/lidar-explorer/#/). They are stored in LAZ format, which is a compressed LiDAR aerial survey (LAS) format. Three tiles were selected and analyzed in this case study.

```
pacman::p_load("rgeoboundaries", "tigris",
"sf", "ggplot2")


library(rgeoboundaries)
library(tigris) #download of shapefiles from the
United States Census Bureau
library(sf)
library(ggplot2)

#We first create a map of the study site from a
larger perspective of the US territory and the
state of Utah. Two sources of the USA border
vector layer can be used.


usa <- gb_adm0("USA")
utah <- gb_adm1("USA")

usa1 <- nation()


#Utah State, where the Dead Horse State Park is
located.


utah1 <- utah %>%
 dplyr::filter(shapeName == "Utah")
```

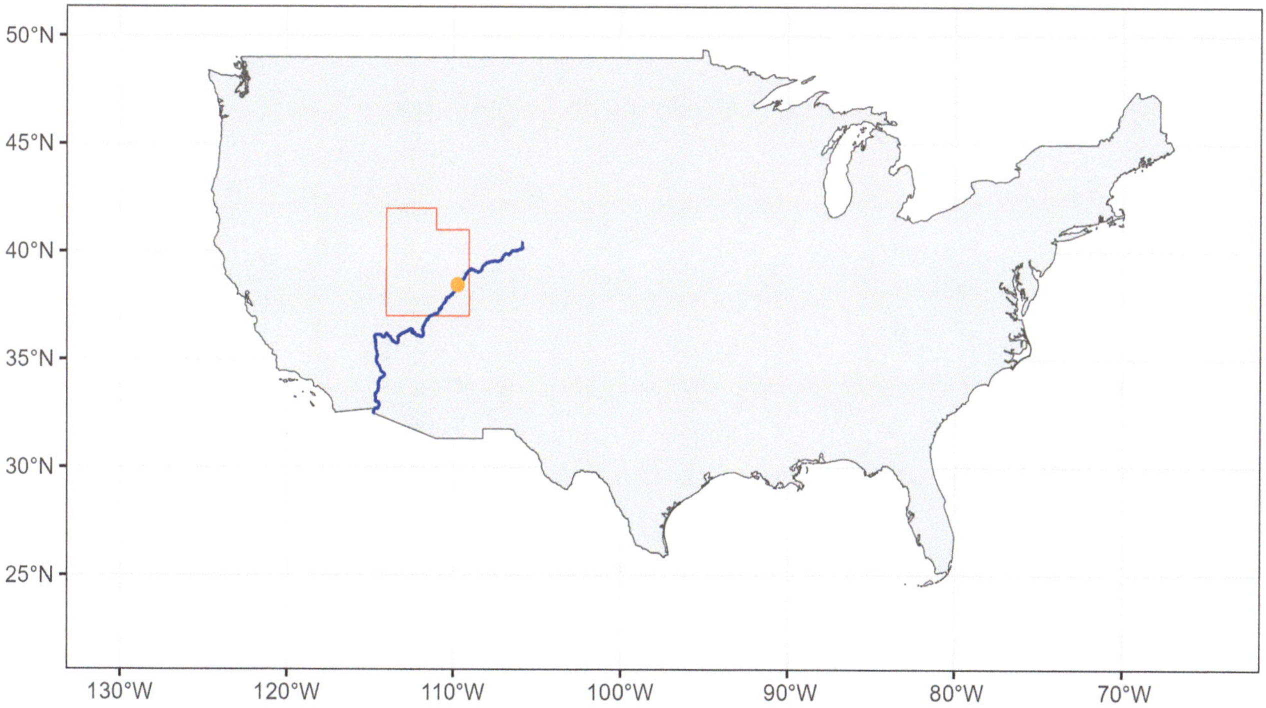

Fig. 11.1 US territory, without Alaska and overseas areas, with the state of Utah (red), the Colorado River (blue), and Dead Horse State Park (orange) shown. The data download process was supported by the **rgeoboundaries** package (Dicko 2024) and the **tigris** package (Walker 2024), the river shapefile from the National Weather Service was used (https://www.weather.gov/gis/Rivers), and the position of the study area was based on Google Maps

```
#US river dataset.
#https://www.weather.gov/gis/Rivers

#URL to the dataset.
url_rivers <- "https://www.weather.gov/source/
gis/Shapefiles/Misc/rv16my07.zip"

#Create a tempfile.
rivers_tempfile <- tempfile(pattern = "rv16my07",
 tmpdir = tempdir(),
 fileext = ".zip")

#Download the file and keep it in a temporary
file.
download.file(url_rivers,
 destfile = rivers_tempfile,
 method = "libcurl")

#Unzip the file.
rivers_unzip <- unzip(rivers_tempfile)

#Read the shapefile using the sf package.
rivers_sf <- read_sf("rv16my07.shp")

#Select only the Colorado River. Why do we use
702 TERMID?
colorado <- rivers_sf %>%
  dplyr::filter(PNAME == "COLORADO R" & TERMID
== 702)

#Approximate geographic position of the study
site.
dh_point <- data.frame(x=-109.75601305888998,
y=38.46038231120897)

#Transform it to a georeferenced point.
dh_point1 <- st_as_sf(dh_point, coords =
c("x","y"), crs = 4326)

Plot the data.
ggplot()+
  geom_sf(data = usa1)+
  geom_sf(data = utah1, color = "red")+
  geom_sf(data = colorado, color = "blue")+
  geom_sf(data = dh_point1, size = 2, color =
"orange")+
  xlim(130,65)+
  ylim(22,50)+
  theme_bw()

#Save the map.
#ggsave("us_map.png")
```

Fig. 11.2 Images of the study area: Google Earth view of two meanders of the Colorado River flowing through Dead Horse State Park (upper panel) and ground view of the Colorado River valley and its steep sides (lower panel). The lower image was taken by Teresa Pawlik (reproduced with permission from the author)

11.2 R in Practice: Point Cloud Data Analysis

For this example, we use the **lidR** package (Roussel et al. 2020). **lidR** is a point-cloud-oriented R package that can be easily integrated with other spatial tools implemented in R (Bivand and Nowosad 2024). To date, it is the only R package dedicated to specialized data such as LiDAR point clouds.

The LiDAR point clouds used here contain points that were already classified according to the standard of the American Society for Photogrammetry and Remote Sensing (ASPRS). Table 11.1 contains the classes that are adopted by the ASPRS (Graham 2012). For our task, points classified as ground (class 2) are needed. Note that some datasets require classification if it is not performed by the data provider or if there are many instances of incorrectly classified points. The study area lacks vegetation, so the ground surface should be correctly reflected in the data. However, there are also many unclassified points highlighted in yellow in Fig. 11.3. As shown in Table 11.2, most of the points, over 3.5 mln, were classified as ground points.

Table 11.1 The standard for classifying points recorded during LiDAR scanning was proposed by the American Society for Photogrammetry and Remote Sensing (ASPRS)

Classification_value	Meaning
0	Created, never classified
1	Unclassified
2	Ground
3	Low vegetation
4	Medium vegetation
5	High vegetation
6	Building
7	Low point (noise)
8	Reserved
9	Water
10	Rail
11	Road surface
12	Reserved
13	Wire - guard (shield)
14	Wire - conductor (phase)
15	Transmission tower
16	Wire-structure connector
17	Bridge deck
18	High noise
19–63	Reserved
64–255	User definable

```
pacman::p_load("lidR", "rgl", "gt", "dplyr")

#The table is based on https://www.asprs.org/
wp-content/uploads/2010/12/LAS_Specification.pdf

asprs_class <- data.frame(Classification_value =
c(0, 1:18, "19-63", "64-255"),
        Meaning = c("Created, never classified",
                    "Unclassified",
                    "Ground",
                    "Low Vegetation",
                    "Medium Vegetation",
                    "High Vegetation",
                    "Building",
                    "Low Point (noise)",
                    "Reserved",
                    "Water",
                    "Rail",
                    "Road Surface",
                    "Reserved",
                    "Wire - Guard (Shield)",
                    "Wire - Conductor (Phase)",
                    "Transmission Tower",
                    "Wire-structure Connector",
                    "Bridge Deck",
                    "High Noise",
                    "Reserved",
                    "User definable"))
```

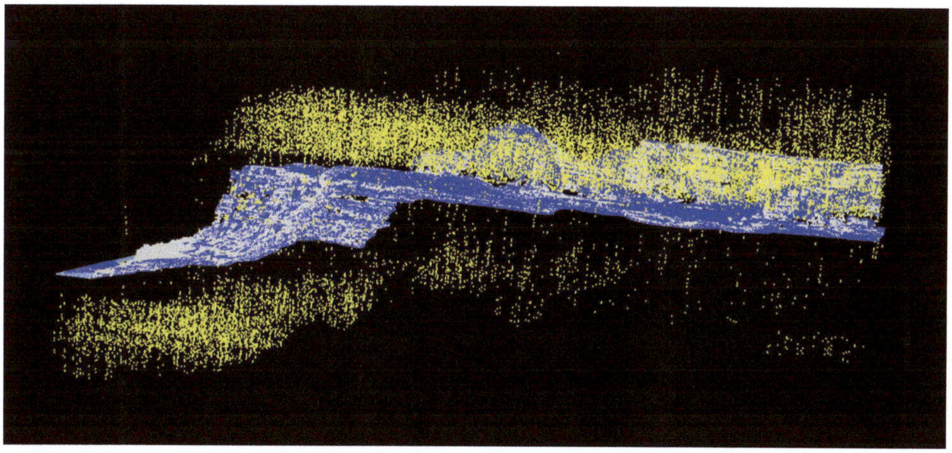

Fig. 11.3 Classified points for the first data tile used in the analysis

Table 11.2 Number of points in the LAS file for the first tile used in the analysis

Classification value	Number of points	Meaning
1	1,176,776	Unclassified
2	3,634,159	Ground
7	9722	Low point (noise)
9	17,460	Water
18	22,496	High noise
20	2457	NA

```
library(gt)


#We modified the table and saved it as a PNG
image.
asprs_tb <- data.frame(asprs_class) %>%
  gt() %>%
  cols_width(everything() ~ px(150)) %>%
  cols_align(align = "center")
asprs_tb

#gtsave(asprs_tb, "asprs_table.png")


 library(lidR)


#Three tiles from the USGS 3DEP LidarExplorer
(https://apps.nationalmap.gov/
lidar-explorer/#/)


url0957 <- "https://rockyweb.usgs.gov/
vdelivery/Datasets/Staged/Elevation/LPC/
Projects/UT_WestEast_B22/UT_WestEast_4_B22/LAZ/
USGS_LPC_UT_WestEast_B22_12SXH0957.laz"
url0857 <- "https://rockyweb.usgs.gov/
vdelivery/Datasets/Staged/Elevation/LPC/
Projects/UT_WestEast_B22/UT_WestEast_4_B22/LAZ/
USGS_LPC_UT_WestEast_B22_12SXH0857.laz"
url0757 <- "https://rockyweb.usgs.gov/
vdelivery/Datasets/Staged/Elevation/LPC/
Projects/UT_WestEast_B22/UT_WestEast_4_B22/LAZ/
USGS_LPC_UT_WestEast_B22_12SXH0757.laz"

#Download the LAZ tiles to your temporary
working directory.
#Define temporary files.

destfile957
<- tempfile(pattern = "USGS_LPC_UT_WestEast_
B22_12SXH0957",
 tmpdir = tempdir(),
 fileext = ".laz")
destfile857
<- tempfile(pattern = "USGS_LPC_UT_WestEast_
B22_12SXH0857",
 tmpdir = tempdir(),
 fileext = ".laz")
destfile757
<- tempfile(pattern = "USGS_LPC_UT_WestEast_
B22_12SXH0757",
                     tmpdir = tempdir(),
                     fileext = ".laz")

#Download the files. Note that it can take some
time.
download.file(url0957, destfile957, mode = "wb")
download.file(url0857, destfile857, mode = "wb")
download.file(url0757, destfile757, mode = "wb")

#Import it to the session memory using the
readLAS() function.
deadhorse1 <- readLAS(destfile957, select =
"xyzc")
deadhorse2 <- readLAS(destfile857, select =
"xyzc")
deadhorse3 <- readLAS(destfile757, select =
"xyzc")

#With the rgl package, point clouds and other
spatial data can be displayed in a new pop-out
window as 3D interactive graphics.


library(rgl)


#Uncheck and run the plot() function if you
want to display the result. Note that plot
rendering can take some time.


#plot(deadhorse1, color = "Classification")
#rgl.snapshot("dh_point_cloud2.png")

#plot(deadhorse1[deadhorse1$Classification ==
2], color = "Classification", bg = "white")
#rgl.snapshot("dh_point_cloud_ground_zoomed_
in1.png")
```

The next figure shows only points classified as ground (Fig. 11.4). The footprint of the scanner device is clearly shown with lines, and some areas are not classified as ground (white patches). Table 11.2 shows that many points were unclassified or classified as noise.

The first tile contains the LiDAR point cloud with points classified into classes 7 (low point, noise), 1 (unclassified), 2 (ground), 18 (high noise), 9 (water), and 20 (reserved). This

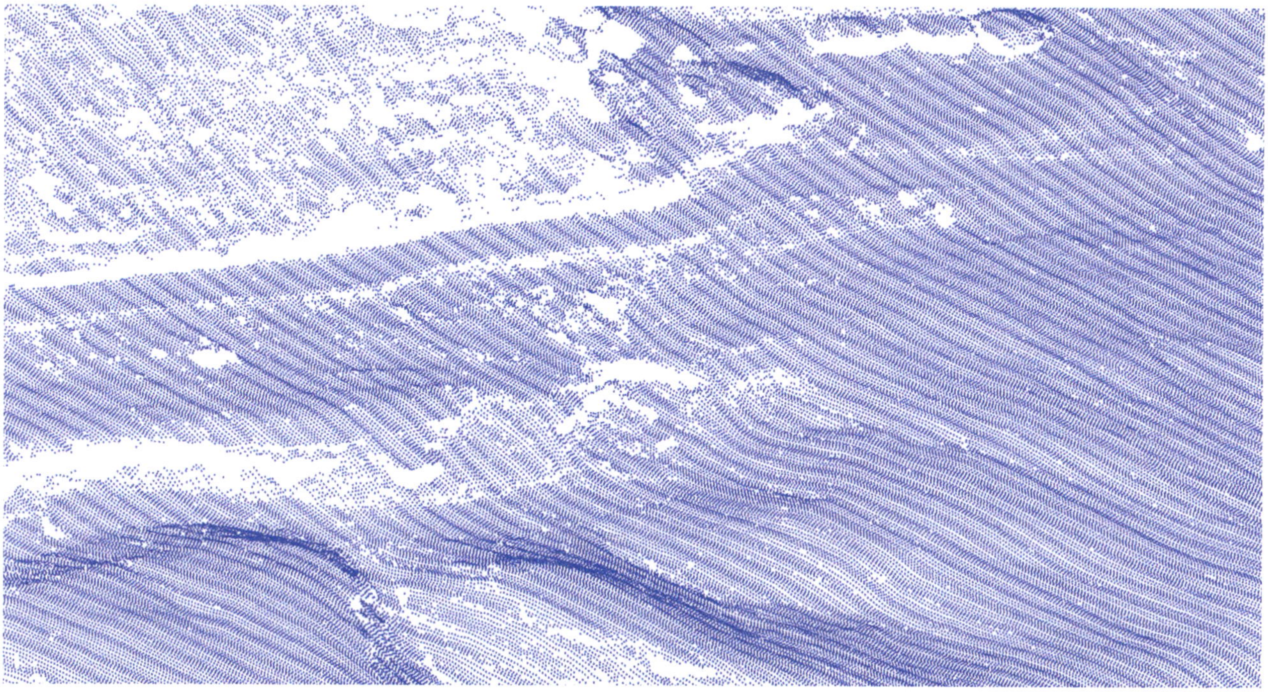

Fig. 11.4 The set of points classified as ground. The data represent a fraction of the analyzed Dead Horse State Park LiDAR point cloud

result can be tested with the ***unique()*** function. However, we wanted to perform an in-depth analysis and determine how many points belong to each class. Notably, we wanted to automate the process and implement it for different datasets. To achieve this, a loop was used inside a custom function. Because the initial number of classes and their types are unknown, we applied the ***unique()*** function to save this information as a separate object. However, note that this object was created only within the function. This information was used in a loop to iteratively update the elements with the ***npoints()*** function from the **lidR** package.

```
#We will check the number of points for each
class (if present in the dataset).


library(dplyr)


#Check deadhorse1 object unique classes.
dh1_unique <-  unique(deadhorse1$Classification)

#We will use a loop inside a customized
function to calculate it for each class and
object. Note that you can insert comments
within functions. The text after # is ignored
during function definition and execution.

function_dh <- function(x){
 dh1_unique <-  unique(x$Classification)
 n_points_list <- list() #empty list

#A loop that helps to automate the analysis.
for(i in seq_along(dh1_unique)){
 n_points <- npoints(x[x$Classification ==
dh1_unique[i]])
 n_points_list[i] <- n_points
}

#A new data frame with information on the
number of points in each class.
dh_df <- data.frame(Classification_value =
as.character(dh1_unique), number = do.
call(rbind, n_points_list))

dh_df1 <- dplyr::left_join(dh_df, asprs_class,
by = "Classification_value")
 return(dh_df1)
}

#Apply the function to the first dataset and
keep the result as an object.
dhsp1_tb <- function_dh(deadhorse1)

dhsp1_tb1 <- data.frame(Classification_value =
as.integer(dhsp1_tb$Classification_value),
                        dhsp1_tb[,-1])
```

```
#Arrange rows following the class label
expressed as an integer.
dhsp1_tb2 <- dhsp1_tb1 %>%
  dplyr::arrange(Classification_value)

#Now, we modify column names.
colnames(dhsp1_tb2) <- c("Classification value",
"Number of points", "Meaning")

#Save the table as an image.
dhsp1_gt <- dhsp1_tb2 %>%
    gt() %>%
  cols_width(everything() ~ px(150)) %>%
  cols_align(align = "center")
dhsp1_gt

#gtsave(dhsp1_gt, "class_point_dhsp1.png")
```

After selecting points classified as ground, the ***rasterize_terrain()*** function is used to obtain a graphical representation of the ground surface. Points classified as ground are subject to interpolation, which is a mathematical transformation of discrete data (points) into continuous data in a grid form (raster layer). In this specific example, the k-nearest neighbor (KNN) method with inverse-distance weighting (IDW) was used. The resulting objects are SpatRasters, a type of object used when working with the **terra** package (Hijmans 2023). Three models representing the ground surface are the subjects of further analyses. First, they were combined into one model using the ***terra::mosaic()*** function (Fig. 11.5). After this step, we can create 2D and 3D models using the interactive features obtained with the ***lidR::plot_dtm3d()*** function (Figs. 11.6 and 11.7).

```
pacman::p_load("terra")

#SpatRaster class (package "terra").

deadhorse1_idw <- rasterize_terrain(deadhorse1[
deadhorse1$Classification == 2],
 algorithm = knnidw(k = 10L, p = 2),
 res = 0.25)
deadhorse2_idw <- rasterize_terrain(deadhorse2[
deadhorse2$Classification == 2],
 algorithm = knnidw(k = 10L, p = 2),
 res = 0.25)
deadhorse3_idw <- rasterize_terrain(deadhorse3[
deadhorse3$Classification == 2],
 algorithm = knnidw(k = 10L, p = 2),
 res = 0.25)

#You can plot the first tile as a rasterized
terrain surface. The plot_dtm3d() function
allows interactive plotting.
#plot_dtm3d(deadhorse1_idw, bg = "white",
legend = TRUE)

library(terra)

#Combain three models into one image.
dh_all <- mosaic(deadhorse1_idw, deadhorse2_
idw, deadhorse3_idw)

#Save the file.
```

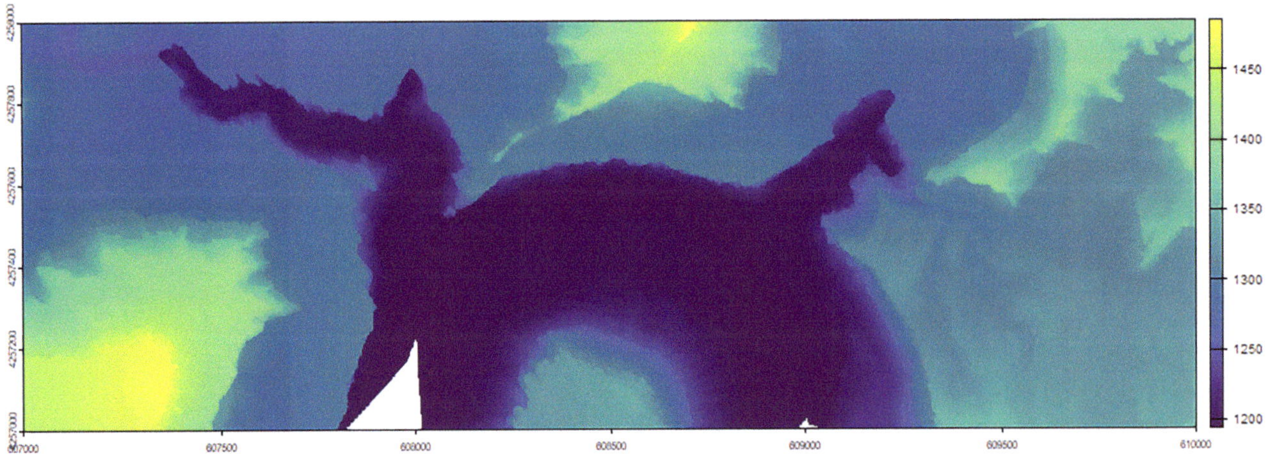

Fig. 11.5 Combined rasterized point clouds with ground points

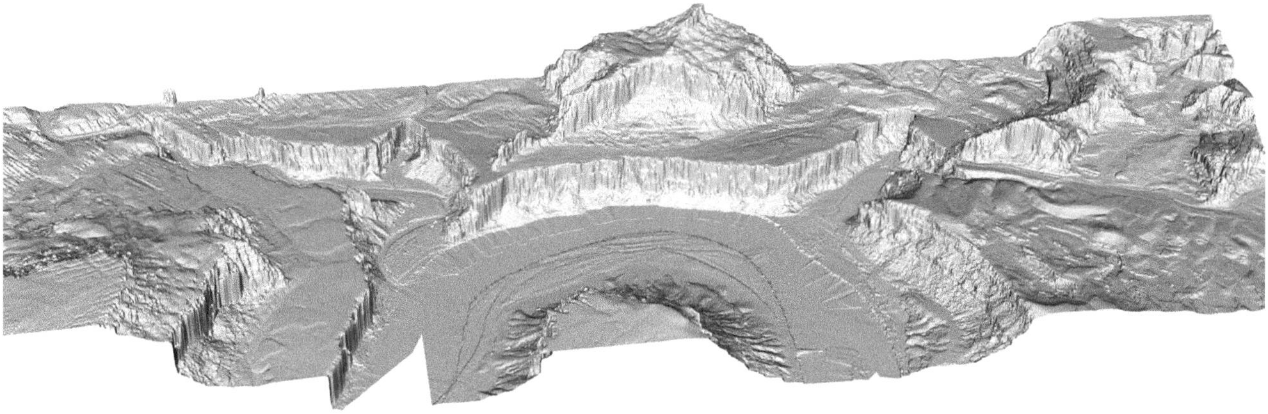

Fig. 11.6 Three tiles with point clouds indicating the ground surface were rasterized and combined into one hillshade image of the digital elevation model

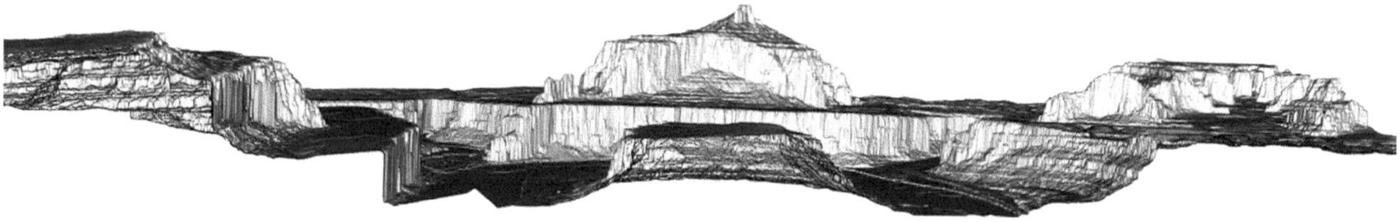

Fig. 11.7 Ground view of the model showing height differences between the valley floor and the highest point in the selected area

```
#png("dem_mosaic1.png", width = 1000, height =
500)
plot(dh_all)
#dev.off()

#Interactive plot in a new pop-out window.
#plot_dtm3d(dh_all, bg = "white")

#After selecting a desired model view, it can
be saved using the rgl.snapshot() function.
#rgl.snapshot("dh_mosaic_hillshade2.png")
```

References

Bivand R, Nowosad J (2024) CRAN task view: analysis of spatial data

Dicko A (2024) rgeoboundaries: geoBoundaries Client, https://CRAN.R-project.org/package=rgeoboundaries

Doelling HH, Chidsey TC Jr (2000) Dead horse point State Park and vicinity geologic road logs, grand and San Juan counties, Utah. In: Geologic road, trail, and Lake guides to Utah's parks and monuments. Utah Geological Association Publication, Salt Lake City, pp 1–38

Dysarz T (2020) Development of methodology for assessment of long-term morphodynamic impact on flood hazard. J Flood Risk Manag 13:e12654. https://doi.org/10.1111/jfr3.12654

Godziek J (2024) Root plates of uprooted trees – automatic detection and biotransport estimation using LiDAR data and field mapping. Int J Appl Earth Obs Geoinf 131:103992. https://doi.org/10.1016/j.jag.2024.103992

Godziek J, Pawlik Ł (2023) Indicators of wind-driven forest disturbances—pit-mound topography, its automatic detection and significance. Catena 221:106757. https://doi.org/10.1016/j.catena.2022.106757

Graham L (2012) The LAS 1.4 specification. Photogramm Eng Remote Sens 78:93–102

Heinaro E, Tanhuanpää T, Yrttimaa T, Holopainen M (2021) Airborne laser scanning reveals large tree trunks on forest floor. For Ecol Manag 491:119225. https://doi.org/10.1016/j.foreco.2021.119225

Hijmans R (2023) terra: spatial data analysis. R Package Version 1(7–1):8

Liu X (2008) Airborne LiDAR for DEM generation: some critical issues. Prog Phys Geogr 32:31–49. https://doi.org/10.1177/0309133308089496

Murtiyoso A, Cabo C, Singh A et al (2024) A review of software solutions to process ground-based point clouds in Forest applications. Curr For Rep 10:401–419. https://doi.org/10.1007/s40725-024-00228-2

NOAA (2012) Lidar 101: an introduction to Lidar technology, data, and applications. National Oceanic and Atmospheric Administration, Charleston, SC

Roussel JR, Auty D, Coops NC et al (2020) lidR: an R package for analysis of airborne laser scanning (ALS) data. Remote Sens Environ 251:112061. https://doi.org/10.1016/j.rse.2020.112061

Walker K (2024) Tigris: load census TIGER/line shapefiles, https://CRAN.R-project.org/package=tigris

Wulder MA, Bater CW, Coops NC et al (2008) The role of LiDAR in sustainable forest management. For Chron 84:807–826. https://doi.org/10.5558/tfc84807-6

Zhen Z, Quackenbush LJ, Zhang L (2016) Trends in automatic individual tree crown detection and delineation—evolution of LiDAR data. Remote Sens 8:333. https://doi.org/10.3390/rs8040333

12 Dynamic Visualization and Animation

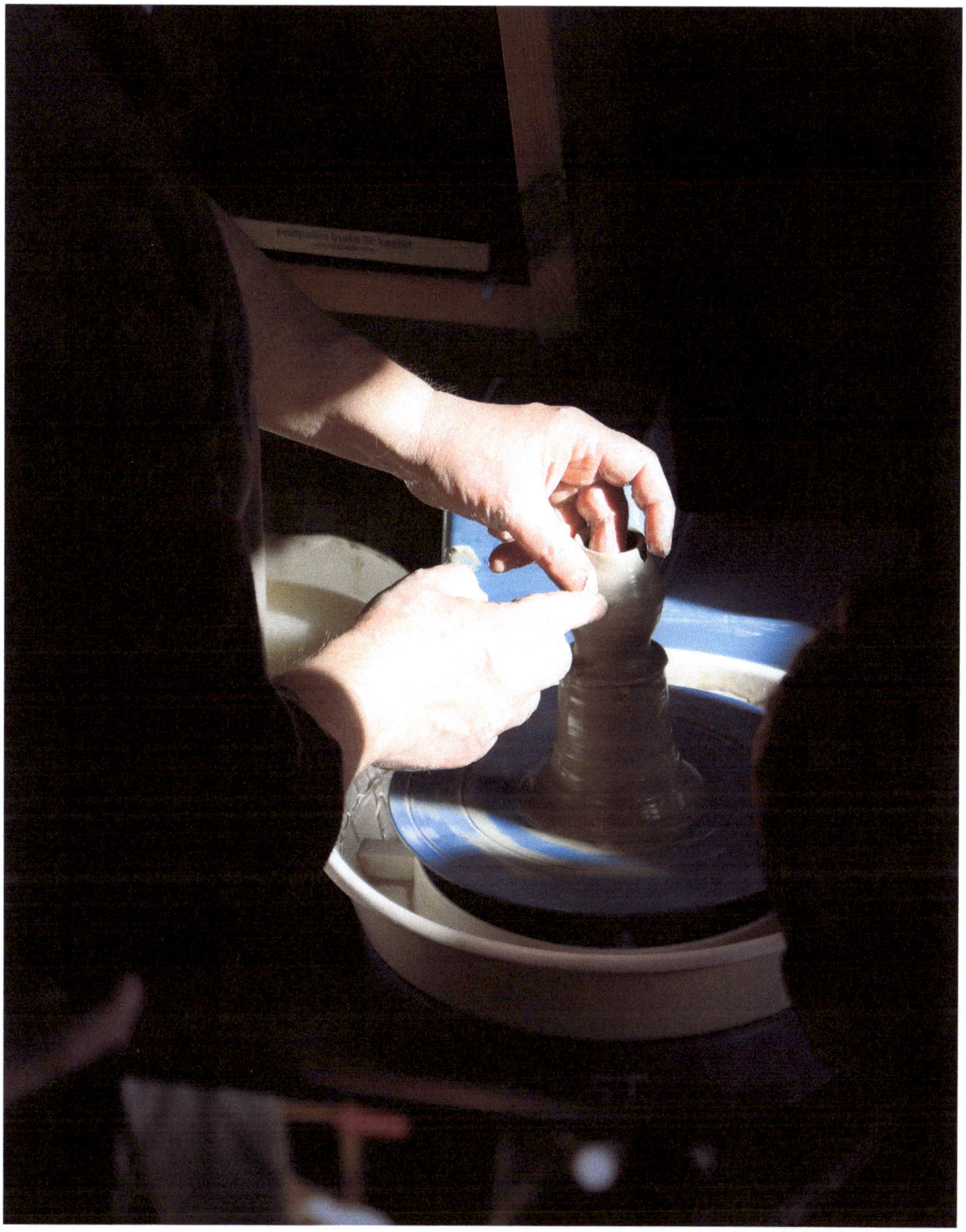

Jadwisin, Poland, 2022, photo: Ł. Pawlik

Supplementary Information The online version contains supplementary material available at https://doi.org/10.1007/978-3-031-89673-6_12.

© The Author(s), under exclusive license to Springer Nature Switzerland AG 2025
Ł. Pawlik, *R Applications in Earth Sciences*, Springer Textbooks in Earth Sciences, Geography and Environment,
https://doi.org/10.1007/978-3-031-89673-6_12

Ever-increasing amounts of data are out there, waiting for interpretation.

Sellars et al. (2013)

Data visualization is one of the most important steps in data mining and modeling (Keim 2001; Sellars et al. 2013; Morse et al. 2017). Science has evolved toward computer- and data-based methods, but new tools and methods are still needed to better capture, analyze, model, and visualize scientific information (Fay 2009). Some tools, techniques, and approaches that were standard in data analysis and presentation several decades or even years ago now need to be updated and improved to support better interpretation of the results and effective communication with the scientific community and the society. In addition, new models and algorithms require high-level graphics and visualization tools to illustrate results in a powerful way. For example, tools that help interpret the results of machine learning and deep learning models, which are frequently based on big data and highly complicated algorithms, are needed. Additionally, with novel approaches, data visualization literacy (DVL), which is the ability to confidently use a given data visualization technique and to extract information from data visualizations, must be evaluated (Börner et al. 2019). It is a common opinion that with the sophisticated tools at hand, a data analyst can easily obtain results, but the key problem is presenting those new results to the public. This is a question related to the efficiency of the message delivery and information flow between scientists and professionals to the end user or audience. This step needs to be as effective as possible because the majority of research is conducted with public funds (which are expected to be returned as new discoveries and information), and the end-user group encompasses nonprofessionals or specialists in other branches of science and industry. Moreover, it is necessary to visualize the outcomes in an easy-to-absorb form to increase their practical value and promote further developments in the respective field.

There are several dynamic and interactive ways in which Earth science-related data and information specific to other branches of science can be presented and displayed. They include animated charts and maps, diagrams with pop-up information on the basis of cursor movement, maps that allow navigation and changes in scale, and 3D diagrams that can be moved in every direction. This list will continue to expand soon.

We learn from cartography that the presentation of graphical information can be as powerful as other methods of scientific and business argumentation. Specifically, *"A map is a special example of a graphic work of art"* (Medyńska-Gulij and Żuchowski 2018). Maps can encompass a massive amount of information, and when designed in an effective manner, they can serve as effective tools to help disseminate scientific discoveries (Medyńska-Gulij et al. 2021). Maps are sometimes called weapons because they can help to gain military and political advantages. Another example includes dynamic web map services that are gates to a variety of spatial information, often extending to the global scale. With the growing number of open spatial data repositories, scientists can search for new discoveries, models, and information. Some classic examples include Google Earth Pro and EarthExplorer, which are provided by the USGS (https://earthexplorer.usgs.gov/). In Europe, the OpenLandMap project (https://openlandmap.org) is under active development, and at the national level, for example, in Poland, the Geoportal spatial data service (https://www.geoportal.gov.pl/en/) has been launched.

12.1 Description of the Data and Project Outline

In the first example, we aim to animate climate time series, which are very interesting datasets because they can feature seasonal and cyclic components and sometimes long-term trends. The seasonal component fluctuates throughout the year and, in most cases, can be easily detected in climate data, especially when observations are from regions with distinct warm and cold seasons. Typically, mean daily records contain several variables that are important from the viewpoint of weather forecasting and modeling. However, for the present task, we select the surface temperature, which is conventionally measured 2 m above the ground surface. The data are from two synoptic stations in Poland: Kasprowy Wierch (1990 m a.s.l.) and Śnieżka (1603 m a.s.l.) (Fig. 12.1).

In the second example, we use rasterized climate observations obtained via spatial modeling. They include monthly data for Norway, Sweden, and Finland. The data were downloaded from the CHELSA database (Karger et al. 2017; Karger and Zimmermann 2018). The slice of these data is used to show how spatial data can be incorporated into an interactive visualization using a dedicated R package.

12.2 R in Practice: Dynamic Plots and Data Animations

Several tools can help a user convert a final product into an animation. For the first example, we use the **gganimate** package (Pedersen and Robinson 2024). The data are downloaded from the archive of the Polish Institute of Meteorology and Water Management using a dedicated R package called **climate** (Czernecki et al. 2020). Daily observations from 2010–2020 for two synoptic stations are considered. We compare data from the two highest-elevation stations in

Fig. 12.1 Images of two synoptic stations where the daily observations used in this task were obtained. These stations are the two highest (elevation) synoptic stations in Poland. Photos by: Łukasz Pawlik

Poland: Kasprowy Wierch (1990 m a.s.l.) and Śnieżka (1603 m a.s.l.). The Kasprowy Wierch synoptic station is in the Tatry Mountains, a region in the Outer Western Carpathians. The Śnieżka synoptic station is in the Karkonosze Mountains, a mountain massif located in the Sudety Mountains, SW Poland. Clear seasonality in these datasets is visible, with low temperatures during the winter season and high temperatures during the summer season (Fig. 12.2).

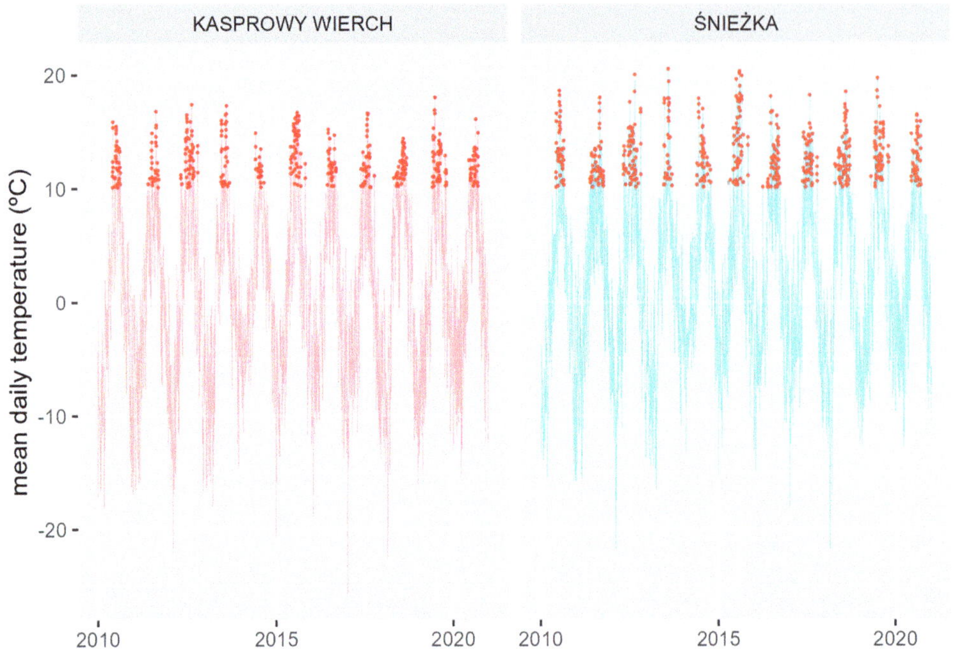

Fig. 12.2 Mean daily temperature at two synoptic stations in Poland between 2010 and 2020: Kasprowy Wierch and Śnieżka. The red dots indicate the mean daily temperature over 10 °C. The time series can be animated using the **gganimate** package

```
pacman::p_load("gganimate", "climate", "dplyr")


library(gganimate)#animations
library(climate)#climate data download

#First, we get data from two synoptic stations
in Poland: Kasprowy Wierch and Śnieżka. The
first is located in the Tatry Mountains, and the
second is in the Karkonosze Mountains.  We are
interested in the daily temperature
observations between 2010 and 2020.


kasprowy <- meteo_imgw(interval = "daily", year
= 2010:2020, station = "KASPROWY WIERCH")
sniezka <- meteo_imgw(interval = "daily", year
= 2010:2020, station = "ŚNIEŻKA")

#Two datasets are modified and merged into one
table. It will make plotting easier.


library(dplyr)


kasprowy1 <- kasprowy %>%
  dplyr::select(3:6, 9)
sniezka1 <- sniezka %>%
  dplyr::select(3:6, 9)
```

```
ks <- rbind(kasprowy1, sniezka1)


#When plotting a time series, a column with
dates is necessary. To have it, we created a
new column with combined information from three
columns (yy, mmm, day) and converted the
character type of data into dates.


ks1 <- ks %>%
  mutate(date = as.Date(paste(yy, mm, day, sep
= "-")))

#When using the notation "package_
name::function_name()", you do not need to call
a package using the library() function.


ggplot2::ggplot()+
  geom_line(ks1, mapping = aes(x=date, y=t2m_
mean_daily, color = station), size = 0.1, alpha
= 0.5)+
  geom_point(ks1[ks1$t2m_mean_daily>10,],
mapping = aes(x=date, y=t2m_mean_daily), color
= "red", size = 0.1)+
  theme
(legend.position = "none", axis.title.x =
element_blank())+
  ylab("mean daily temperature (°C)")+
```

```
  facet_wrap(~station)

#Save the plot.
#ggsave("climate_ts.png", width = 15, height =
10, units = "cm")

#Create animation.
anim_ks <- ggplot()+
  geom_line(ks1, mapping = aes(x=date, y=t2m_
mean_daily, color = station))+
  geom_point(ks1[ks1$t2m_mean_daily>10,],
mapping = aes(x=date, y=t2m_mean_daily), color
= "red")+
  theme
(legend.position = "none", axis.title.x =
element_blank())+
  ylab("mean daily temperature (°C)")+
  facet_wrap(~station)+
  transition_reveal(date)+
  ease_aes('linear')
anim_ks

#animate(anim_ks, renderer = gifski_renderer())
#Save the animation.
#anim_save("anim2.gif")
```

For the second example, we use tools available in the **tmap** package (Tennekes 2018). The ***tmap_animate()*** function is used to animate changes in surface temperature across Norway, Sweden, and Finland. Before we can obtain the final animated map, the data need to be downloaded and saved in a temporary directory as SpatRaster objects. Here, we work with the CHELSA datasets (Karger et al. 2017; Karger and Zimmermann 2018). These data can be downloaded with the **climenv** package (Tsakalos et al. 2023) or the **ClimDatDownloadR** package (Jentsch et al. 2024). However, we introduce a custom function to show how to work with functions, loops, and temporary files. First, two sets of temporary file names and URL links are created to download individual files to a temporary memory. With those two lists, files were downloaded iteratively from the CHELSA web service with a loop. Before importing the next file, the previous file was cropped to the spatial range of Norway, and the original file was removed from the memory space. Next, the custom function was created using *function()* with two arguments: the index from the list of subsequent months and years and the country name to which the boundary of the data is cropped. To reduce the size of the temporary file, the country name was changed to the ISO3 code using the **countrycode** package (Arel-Bundock et al. 2018). Using the ***get_chelsa_tmax()*** function in a loop, we obtained a spatiotemporal series of monthly data for 2 years. The next step is to use those files and create a series of **tmap** objects that are used in the ***tmap_animation()*** function. Importantly, all CHELSA products are in a geographic coordinate system referenced to the WGS84 horizontal datum. Some CHELSA temperature data are stored in Kelvin * 10 units. For this reason, to obtain the temperature in degrees Celsius, values must be transformed such that T[°C] = X/10–273.15 (Khaydarov and Gerlitz 2019). In our example, we use the maximum daily temperature averaged over the month. The temperature is in degrees Celsius multiplied by 10. For this reason, before constructing a map, the temperature values were divided by 10 to restore the real values. The first map displays only a slice of the data for January 2008 (Fig. 12.3). The second figure shows the spatiotemporal array of monthly temperatures in a facet grid form, with a matrix of panels defined on the basis of row and column variables (Fig. 12.4).

```
pacman::p_load("sf", "raster", "terra", "tmap",
"countrycode", "purrr", "stringr")


library(sf)
library(raster)
library(terra)

#In this part, we want to animate monthly
temperature changes over Norway for 2008-2009.


yrs <- 2008:2009
yrs1 <- rep(yrs, each=12)

mth <- 1:12
mth1 <- rep(mth, times = 2009-2008+1)

time_per <- data.frame(year=yrs1, month=mth1)


num_seq <- 1:length(mth1)


#Empty lists used during the loop sequence will
be filled with information on file names and URL
links used for data download.


files_tmax <- list()
url_tmax <- list()

#A loop during which a list of files is created
with a desired combination of months and years.
This information helps to create URL links for
data download from the CHELSA database.
```

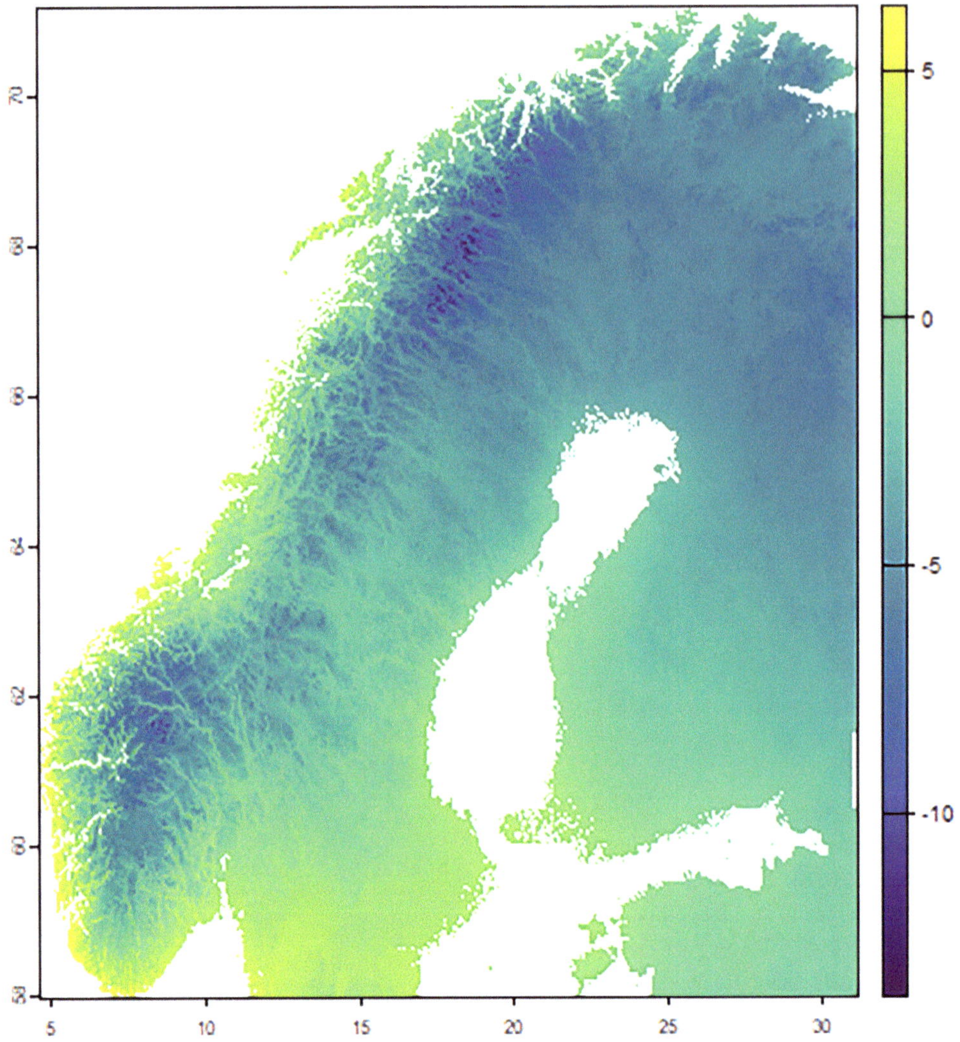

Fig. 12.3 Mean temperature in January 2008 in Scandinavia. Temperatures are in Celsius degrees (°C)

```
for(i in seq_along(num_seq)){
  file_name <- paste0('CHELSAcruts_tmax_',
time_per$month[i], '_', time_per$year[i])
  files_tmax[[i]] <- file_name
}

#The second loop helps to construct URL links
for data download. If the link to the data will
change in the future (we rely here on the
external data provider), the loop below must be
modified.

for(i in seq_along(num_seq)){
  url_address <- paste0('https://os.zhdk.cloud.
switch.ch/chelsav1/chelsa_cruts/tmax/
CHELSAcruts_tmax_',
                          time_per$month[i], '_',
time_per$year[i], '_V.1.0.tif')
  url_tmax[[i]] <- url_address
}

#Merge elements of lists into data frames.
files_tmax1 <- do.call(rbind, files_tmax)
url_tmax1 <- do.call(rbind, url_tmax)

#The next step is using a custom function that
automates data download. Note that each file is
downloaded and kept in a temporary file, then
cropped and saved in a temporary directory. The
original file has been removed from memory.

get_chelsa_tmax <- function(x, country){
  destfile <- tempfile(tmpdir = tempdir(),
                       fileext = ".tif")
  #download files using URLs from the list
  download.file(url_tmax1[x],
```

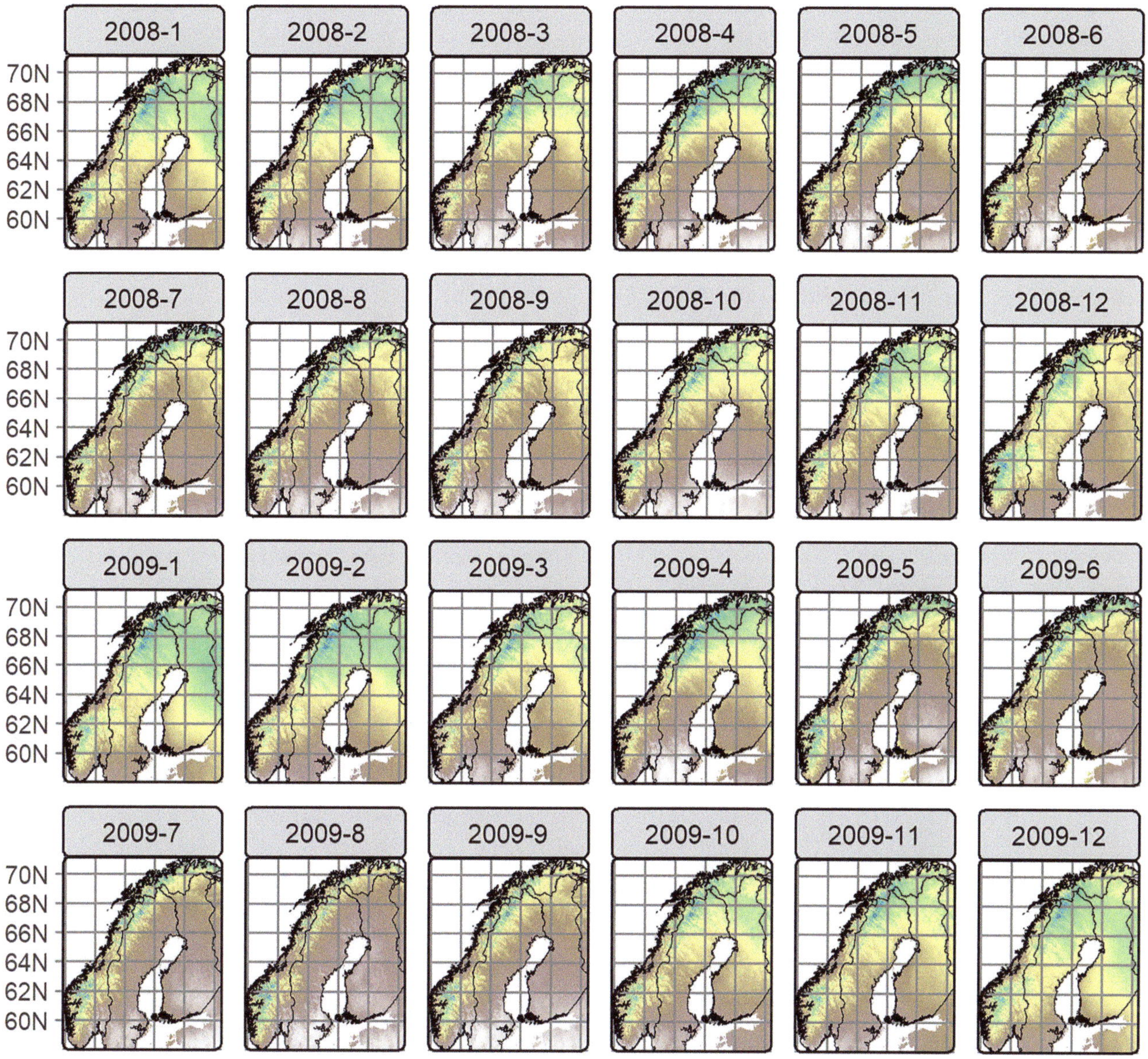

Fig. 12.4 Monthly data from 2008 and 2009 with surface temperature for Norway, Sweden, and Finland from the CHELSA database

```
                destfile = destfile,
                mode = 'wb')
  #download state border of the selected
country
  cntry <- rgeoboundaries::gb_adm0(country)
  #import data as a SpatRaster object
  raster1 <- terra::rast(destfile)
  #construct a temporary file name with
information on data product, date, and country
label
  destfile1 <- tempfile(pattern = paste0(files_
tmax1[x], "crop", countrycode::countrycode(c(co
untry), "country.name", "iso3c")),
    tmpdir = tempdir(), fileext = ".tif")
  #crop the raster and save it in a temporary
memory
  terra::crop(raster1, terra::vect(cntry),
            filename=destfile1)
  #remove a temporary file name
  unlink(paste0(destfile1,'.tif'))
}

#Use a loop and the get_chelsa_tmax() function
to get the data.
for(i in num_seq){
  get_chelsa_tmax(i, "Norway")
}
```

```
#List all files from the temporal directory with
the "cropped" term in their names.
norway_tif <- list.files(tempdir(), pattern =
"cropNOR", full.names = TRUE)

#Inspect file names and, if needed, select only
a part of it.
norway_tif

#We select only 24 files at the top of the list.
Note that you can have many different files in
the current session temporal memory, especially
if you have tried downloading different data for
various countries. For this reason, we
introduced the "cropNOR" pattern to temporary
file names. We also added information from the
files_tmaxl list, which is a product name and
date.


#Note that files are not arranged correctly.
They need to be rearranged to follow months and
years.


#We will now extract information from temporary
file names. This information will help us to
rearrange the files.


library(stringr) #package for operations with
strings (text)


#Construct a data frame with file names, months,
and years
norway_tif_df <- data.frame(link = norway_tif,
                            month = as.
integer(str_extract(str_extract(norway_tif,
"_[0-9]+_"), "[0-9]+")),
                            year = as.
integer(str_extract(norway_tif, "20[0-9]+")))

#Arrange rows according to months and years.
norway_tif_df <- norway_tif_df |>
  dplyr::arrange(month) |>
  dplyr::arrange(year) |>
  dplyr::filter(!is.na(month))

#Inspect the object.
norway_tif_df

#Display only one map. The data is in Celsius
degrees multiplied by 10. The results need to
be divided by 10 to get the final result. It can
be done using the raster algebra.
rast1 <- terra::rast(norway_tif_df[1,1])/10


#Plot the file. If you want to save the map,
please uncheck the png() and dev.off() functions
and modify the file directory.


#png("nor200801.png")
plot(rast1)
#dev.off()

#Create a raster stack with all raster layers.
rast_all <- c(rast(norway_tif_df[,1]))

library(tmap)
library(purrr)

#Create temporary file names for tmap objects.
file_number <- data.frame(prx = "tmap",
                          number =
1:length(norway_tif_df[,1]))

tmap_file_names <- purrr::map2(file_number$prx,
file_number$number, paste0)


#Empty list for data storing.
list_tm_objects <- list()

for(i in 1:length(tmap_file_names)){
  Norway <- rgeoboundaries::gb_adm0("Norway")
  Sweden <- rgeoboundaries::gb_adm0("Sweden")
  Finland <- rgeoboundaries::gb_adm0("Finland")

  years <- 2008:2009
  years1 <- rep(years, each=12)

  months <- 1:12
  months1 <- rep(months, times = 2009-2008+1)

  time_seq <- data.frame(year=years1,
month=months1)


  tmap_obj <- tm_shape((rast_all[[i]]/10))+
    tm_raster(col.scale = tm_scale_
continuous(values = "matplotlib.terrain"),
              col.legend = tm_legend(title =
paste(time_seq$year[i], time_seq$month[i], sep
= "-"),
                                   title.size
= 1))+
    tm_legend(reverse = TRUE)+
    tm_graticules()+
    tm_shape(Norway)+
```

```
    tm_lines()+
    tm_shape(Sweden)+
    tm_lines()+
    tm_shape(Finland)+
    tm_lines()

  list_tm_objects[[i]] <- tmap_obj
}

#If you want to save the animation in your
working directory, use the filename argument.
tmap_animation(list_tm_objects, width = 600,
height = 600, delay = 60)

#Static maps displayed in a facet grid.


Norway <- rgeoboundaries::gb_adm0("Norway")
Sweden <- rgeoboundaries::gb_adm0("Sweden")
Finland <- rgeoboundaries::gb_adm0("Finland")

#With too many facets, displaying this
collection of maps can take too long. In this
case, we should consider reducing the number of
facets to 12.


tmap_obj_facet <- tm_shape(rast_all[[1:12]])+
  tm_raster(col.scale = tm_scale_
continuous(values = "matplotlib.terrain"))+
  tm_layout(panel.labels = paste(time_seq$year,
time_seq$month, sep = "-"),
            legend.show = FALSE)+
  tm_legend(reverse = TRUE)+
  tm_graticules()+
  tm_shape(Norway)+
  tm_lines()+
  tm_shape(Sweden)+
  tm_lines()+
  tm_shape(Finland)+
  tm_lines()+
  tm_facets_grid()

#Display the map
tmap_obj_facet

#Save the figure.
#tmap_save(tmap_obj_facet, "norway_facets.png",
width = 1600, height = 1400, units = "px")
```

Interactive maps are another example of dynamic data visualization using open data and tools. The map in Fig. 12.5 was created with the **leaflet** package (Cheng et al. 2023). It is based on the Leaflet JavaScript library and the **htmlwidgets** package (Vaidyanathan et al. 2023). As a raster layer, we use the mean monthly temperature for July 2008 downloaded from the CHELSA database. The **leaflet** package allows maps to be zoomed in and out and changes the viewing perspective to support navigation to other places on the digital globe. With this interactive approach, we can deliver more information than when we use a static map and show the data in a more interesting way.

```
pacman::p_load("leaflet", "htmlwidgets",
"webshot2", "mapview")


library(leaflet)
library(mapview)

#Data for July 2008.
rast2 <- terra::rast(norway_tif[7,1])/10

#We can save the file and use it later.
terra::writeRaster(rast2,"chelsaJul2008.tif")

#Definition of the color scale used for the
raster layer.
pal <- colorNumeric(c("#0C2C84", "#41B6C4",
"#FFFFCC"), values(rast2),
 na.color = "transparent")

#Dynamic map using the leaflet package.
map_chelsa <- leaflet() %>%
 addTiles() %>%
 addRasterImage(rast2, colors = pal, opacity =
0.7) %>%
 addLegend(pal=pal, values = values(rast2),
title = "temperature")

#Display the map.
map_chelsa

#Save the map as a PNG file.
library(htmlwidgets)

saveWidget(map_chelsa, "leaflet_map.html",
selfcontained = FALSE)


#If you want to save the file, the file directory
in the webshot() function must be specified.
webshot2::webshot("leaflet_map.html", file =
"leaflet_map.png",
 cliprect = "viewport")
```

Fig. 12.5 The raster layer represents the mean monthly temperature in July 2008. The OpenStreetMap was accessed using the **leaflet** package. This map can be viewed in interactive mode

References

Arel-Bundock V, Enevoldsen N, Yetman CJ (2018) Countrycode: an R package to convert country names and country codes. J Open Source Software 3:848. https://doi.org/10.21105/joss.00848

Börner K, Bueckle A, Ginda M (2019) Data visualization literacy: definitions, conceptual frameworks, exercises, and assessments. Proc Natl Acad Sci 116:1857–1864. https://doi.org/10.1073/pnas.1807180116

Cheng J, Schloerke B, Karambelkar B, Xie Y (2023) Leaflet: create interactive web maps with the JavaScript "leaflet" library. R package version 2.2.0

Czernecki B, Głogowski A, Nowosad J (2020) Climate: an R package to access free in-situ meteorological and hydrological datasets for environmental assessment. Sustain For 12:394. https://doi.org/10.3390/su12010394

Fay D (2009) Introduction. In: The fourth paradigm. Data-intensive scientific discovery. Microsoft Research, Redmond, Washington, p 287

Jentsch H, Bobrowski M, Weidinger J (2024) ClimDatDownloadR: downloads climate data from Chelsa and WorldClim

Karger DN, Zimmermann NE (2018) CHELSAcruts—high resolution temperature and precipitation timeseries for the 20th century and beyond. EnviDat. https://doi.org/10.16904/envidat.159

Karger DN, Conrad O, Böhner J et al (2017) Climatologies at high resolution for the earth's land surface areas. Sci Data 4:170122. https://doi.org/10.1038/sdata.2017.122

Keim DA (2001) Visual exploration of large data sets. Commun ACM 44:38–44. https://doi.org/10.1145/381641.381656

Khaydarov M, Gerlitz L (2019) Climate variability and change over Uzbekistan—an analysis based on high resolution CHELSA data. Central Asian J Water Res 5:1–19. https://doi.org/10.29258/CAJWR/2019-R1.v5-2/119.eng

Medyńska-Gulij B, Żuchowski TJ (2018) European topography in eighteenth-century manuscript maps. Bogucki Wydawnictwo Naukowe, Poznań

Medyńska-Gulij B, Forrest D, Cybulski P (2021) Modern cartographic forms of expression: the renaissance of multimedia cartography. ISPRS Int J Geo Inf 10:484. https://doi.org/10.3390/ijgi10070484

Morse P, Reading A, Lueg C (2017) Animated analysis of geoscientific datasets: an interactive graphical application. Comput Geosci 109:87–94. https://doi.org/10.1016/j.cageo.2017.07.006

Pedersen TL, Robinson D (2024) gganimate: a grammar of animated graphics, https://CRAN.R-project.org/package=gganimate

Sellars S, Nguyen P, Chu W et al (2013) Computational earth science: big data transformed into insight. EOS Trans Am Geophys Union 94:277–278. https://doi.org/10.1002/2013EO320001

Tennekes M (2018) tmap: thematic maps in R. J Stat Softw 84:1–39. https://doi.org/10.18637/jss.v084.i06

Tsakalos JL, Smith MR, Luebert F, Mucina L (2023) climenv: download, extract and visualise climatic and elevation data. J Veg Sci 34:1–7. https://doi.org/10.1111/jvs.13215

Vaidyanathan R, Xie Y, Allaire J, et al (2023) htmlwidgets: HTML widgets for R, https://CRAN.Rproject.org/package=htmlwidgets

13 Final Comments, Conclusions, and Data Science Perspective

R offers a well-established statistical and graphical environment for efficient data analysis, modeling, and visualization. Owing to the constantly growing number of packages dedicated to Earth science data exploration and interpretation, high-level solutions can now be obtained for scientific publications, reports, and professional documents. The growing volume of data is a positive outcome of technological advancements. Owing to these improvements, we can better understand our Blue Planet and model and predict many aspects of the global geoecosystem, with a variety of abiotic and biotic components and considerable natural complexity. However, it is also a challenge to obtain knowledge from this massive amount of data, which is frequently inaccurate, incomplete, or difficult to interpret. Moreover, the work of a data analyst requires a detective's precision and scrutiny rather than a public speaker's generalizations. Multidimensional and multivariate datasets are now the norm, and new technological and methodological solutions are urgently needed to analyze them.

In this book, we explored different data types, volumes, and complexities. The list of data types presented is definitely incomplete, and undoubtedly, more work is needed to fully explore the rich variety of geoscientific information and the ways in which they can be analyzed and displayed. The first step, related to R and the Earth science data, included in this book offered an introduction to what we can expect when diving into the immensely intriguing and fascinating world of geoscience and environmental data.

It is true that we cannot stop the growing data currents. We need the data, but we cannot absorb all of this information. Moreover, raw data are useless if not effectively explored, and it can be frustrating when no tools are available for data exploration. Datasets often contain abundant and valuable information that can positively affect our daily lives, for example, by providing information for new drug development to combat a serious disease. This issue drove Dr. John Swallow to map all instances of cholera in reference to the locations of water pumps in central London. When two datasets were combined into one, a valuable answer was obtained.

Open data and open-source software are becoming increasingly common. However, in many countries, no active initiatives are promoting high-precision spatial data production. Such production can be restricted by various factors, ranging from the high cost of such actions to the political decisions that are made regarding open spatial data. We feel that Earth science data or any scientific data that could improve the quality of life and our knowledge of the surrounding world should not be restricted from wider user access. It seems that many years still need to pass before the open data initiative takes hold at the global scale.

R can help us obtain knowledge from data clouds. One benefit of R is data visualization, which has become increasingly important for modern society, with many individuals unable to focus on one issue for too long. With a growing amount of information, we often miss details. With the increasing number of scientific papers and the interdisciplinarity of research teams, effective communication among team members, scientific teams, and the public has become a key issue at every step of the scientific process. If our goal is to safely navigate the data cloud, we need sufficient tools and a deep understanding of methods to make our results easily understandable and self-explanatory to readers. By reaching the end-user, the public, or the reader, we can highlight the real knowledge that can be gained from data and show how powerful the underlying science is.

© The Author(s), under exclusive license to Springer Nature Switzerland AG 2025

Ł. Pawlik, *R Applications in Earth Sciences*, Springer Textbooks in Earth Sciences, Geography and Environment,
https://doi.org/10.1007/978-3-031-89673-6_13

List of Terms

© The Editor(s) (if applicable) and The Author(s), under exclusive license to Springer Nature Switzerland AG 2025

Ł. Pawlik, *R Applications in Earth Sciences*, Springer Textbooks in Earth Sciences, Geography and Environment,
https://doi.org/10.1007/978-3-031-89673-6

List of R Packages

© The Editor(s) (if applicable) and The Author(s), under exclusive license to Springer Nature Switzerland AG 2025

Ł. Pawlik, *R Applications in Earth Sciences*, Springer Textbooks in Earth Sciences, Geography and Environment,
https://doi.org/10.1007/978-3-031-89673-6

MIX
Papier aus verantwortungsvollen Quellen
Paper from responsible sources
FSC® C105338

If you have any concerns about our products,
you can contact us on
ProductSafety@springernature.com

In case Publisher is established outside the EU,
the EU authorized representative is:
Springer Nature Customer Service Center GmbH
Europaplatz 3, 69115 Heidelberg, Germany

Printed by Libri Plureos GmbH
in Hamburg, Germany